Recognizing the Autonomy of Nature

THEORY AND PRACTICE

Thomas Heyd, EDITOR

COLUMBIA UNIVERSITY PRESS
NEW YORK

Columbia University Press
Publishers Since 1893
New York Chichester, West Sussex

Chapter 2 is based on "Nature as Agency and the Prospects for a Progressive Naturalism," *Capitalism, Nature, Socialism* 10, no. 4, issue 48 (2001): 3–32.

Chapter 4 is based on "The Liberation of Humanity and Nature," *Environmental Values* 11, no. 4 (November 2002): 397–405.

Chapter 8 is based on "Purple Loosestrife and the Bounding of Nature in North American Wetlands," *UnderCurrents: Journal of Critical Environmental Studies* 9 (1997).

Library of Congress Cataloging-in-Publication Data

Recognizing the autonomy of nature : theory and practice / edited by Thomas Heyd.
p. cm.
Includes bibliographical references and index.
ISBN 0–231–13606–4 (cloth : alk. paper)
1. Environmental ethics. I Heyd, Thomas, 1956–

GE42.R43 2005
179'.1—dc22 2005045567

Columbia University Press books are printed on permanent and durable acid-free paper.

Printed in the United States of America
c 10 9 8 7 6 5 4 3 2 1

For Elke and Mustafa, and for Karin, at whose homes I have been able to work happily so often

Contents

Preface and Acknowledgments

THE IDEA for this book came about while I was preparing for a conference. My colleague and friend, Antony Berger, a geologist interested in the reception of drastic natural events, such as occur during rapid climate change, had asked me to join him in organizing a multidisciplinary meeting to discuss the ethics of natural environmental change. What he primarily had in mind was the question whether nature could be held morally responsible for the effects of events such as floods or earthquakes.

Although rather skeptical about the sense of holding nature responsible, I was intrigued by the question whether nonhuman nature in its totality, or any part of it, may be said to *act* at all. If it reasonably could be said that it acts, even if only in an attenuated way, it would open up the space to ask whether this entailed moral responsibilities for human beings toward nature and perhaps for nature toward human beings. The result of our deliberations was that we ended up with a compromise title for the conference, calling it "Recognizing the Autonomy of Nature: Ethics and Natural Environmental Change."

The conference, comprising colleagues mostly from philosophy and environmental studies but also from other sciences such as environmental sociology and geology, gathered for a total of eight days at various locations in Newfoundland. The meetings were intended to combine standard academic presentations with a field trip designed to provide participants insight into the actual condition of the natural environment on the island. According to this plan, after meeting at the University of Newfoundland in St. John's, conference participants all moved across to the west coast of the island in order to meet with National Park rangers, specialists in forestry and biological resources, and representatives of outport communities who had hands-on experience with the environmentally as well as economically calamitous collapse of certain fish species, such as cod, off their coasts.

After the event, I decided to pursue the topic of the conference further, focusing primarily on whether it makes sense to speak of the autonomy of nature and the implications of this issue for human practices affecting the environment. A call for papers succeeded in complementing my selection of outstanding papers from the conference with essays from some of the most important environmental philosophers who have dealt with the topic. The result of the combination of these papers is this collection of wonderful essays.

Finally, Columbia University Press urged me to obtain one further essay that could function as a concluding, critical commentary on all those collected. I am very pleased that Bill Jordan, a noted theorist and practitioner of restoration, accepted the invitation and hence provided an appropriate closure for the book.

I would like to acknowledge the help of many friends in getting the original conference organized. Andrew Light and Eric Katz were of extraordinary help in getting the manuscript published. I also would like to thank Wendy Lochner of Columbia University Press for her patience. My greatest gratitude goes to the contributors to the volume for their creative work and their willingness to go through the long editing process with me. I have come to know them as intimate mates on this journey toward a better understanding of our place in relation to the rest of nature. Thanks and good wishes to all.

Victoria, British Columbia

Recognizing the Autonomy of Nature

ONE

Introduction: Recognizing the Autonomy of Nature: Theory and Practice

THOMAS HEYD

IN CONTEMPORARY times, anthropogenic changes in the natural environment have become so considerable that a rethinking of our theory and our practice with regard to nature has become imperative. This book is intended as an exploration of the idea that this rethinking calls for *the recognition of the autonomy of nature,* understood both in epistemological and ethical terms.[1] The book is hence intended to explore, on the one hand, the reasons for attributing autonomy to nature and, on the other, what respect for nature's autonomy practically implies.

In this introduction, I begin with a description of a particular place and its human and nonhuman inhabitants, which incited me to reflect on practically effective proconservation arguments. Next, I clarify how most proconservation arguments can become ineffective because of changes in perspective among human valuers, while arguments that appeal to intrinsic value, which in turn are based on the recognition of the autonomy of nature, have a special place.[2] Then, I briefly discuss how, on the basis of their unique structure, such arguments may be effective even when arguments appealing to self-interest or aesthetics are not. After this, I introduce the papers featured in this book.

A Place and Its Circumstances

The town where I grew up is located on a hill by the sea in a broad alluvial plain crossed by a small river known in ancient times for its health-bringing waters. Although the river now only carries a trickle of a stream as a result of a dam, built fifty years ago, located a day's walk farther up, it continues to nourish a rich vegetation both in its own winding valley and in the adjacent valleys filled with irrigated fields and orange orchards. The beautiful but infamously poisonous oleander (*Nerium oleander*), flowering bright red and white, marks the river's borders.

The river flows into a semicircular bay bordered by two mountains, both a half a day's walk from town, one to the southeast and one to the north. Both of the mountains encounter the Mediterranean Sea with precipitous cliffs. The mountain located on the northern end of the alluvial plain quickly rises up to such heights that in winter its pine trees are sometimes covered with a fine coating of snow. Interrupted by deep canyons, this northern mountain connects with an even higher, extended range of sierras all along the western side of the plain. Though only a few long day walks away, these western ranges are so high that in former times packed snow would survive in caves until summer. Enterprising country folk would load up portions of the resulting ice in late spring to supply the towns on the plain with the means to refrigerate their foods.

The southern end of the plain, bordered by the relatively low mountain to the southeast, also meets the sea in the south, a day's walk away from town. If it were not for the two openings to the sea and the narrow strip of desertic, hilly land that can be traversed toward the southwest permitting access further along the coast, this plain would be quite isolated from the rest of the world.

Nearly the entirety of the plain is dotted with rounded, pine-clad hillocks. The older towns sit on craggy outcrops, traditionally watchful for slaving pirate ships out for booty. Newer towns nestle in the valleys, closer to the orange and *níspero* orchards in the low irrigated areas and the olive, fig, carob, and almond tree groves in the higher, dryland parts. Despite the dangers that the area had to face in the past, the plain exudes a sense of well-being and safety. In the past, no great riches could be accumulated here, but with the sophisticated irrigation techniques, introduced by the Romans during their centuries-long occupation and perfected by people from Northern Africa and Arabia thereafter, the area could offer its inhabitants all the varied food typical of the Mediterranean basin. A great Muslim scholar is supposed to have lived high up in one of the narrow mountain valleys, which indicates that the area was flourishing in more than agricultural ways.

What is most interesting from the point of view of environmental philosophy is that the natural and the artifactual merge and coexist throughout this area at uncounted places. Irrigated citrus orchards may suddenly give way to a hill covered with wild pine trees and dryland herbs, such as rosemary, basil, and thyme. On the other side of the hill, however, there may be an olive tree grove, itself perhaps bordered by the wild canyons of one of the mountains that enclose the area. High up on the northern mountain there are again a few almond groves, since they do not require irrigation, alternating with pine forests and the odd shepherd's hut, built with local stones and mud. In the lowlands, one can encounter families of wild pigs scurrying across a road to the next creek filled with giant reed (*Arundo donax*) and oleander. The seashore along the bay is covered with rounded boulders, which have come rolling down the river valley throughout thousands of years. Since human inhabitation, those boulder beaches have served as relatively stable barriers against the encroachments of heavy winter storms on fields nearby.

Over the last forty years, however, this whole area has undergone an increasingly radical transformation. Where the plain meets the sea along its southern edge, a city of skyscrapers has been built for sun-hungry northern Europeans. Throughout the plain, residential homes for summer visitors and retirees have been filling in all sites that have a view toward the sea. My hometown has started to spread along the seashore. Houses are even being built on the steep slopes of the two mountains that border the bay. Presently, new possibilities of growth are being considered by developers and city councils. Two of the harbors, mostly housing yachts, are to be expanded, one of them destroying a sizable field of *posidonia oceanica* of crucial importance for the diversity of fish species that it provides a home for. A new, fast highway has been built into the western mountain ranges, making urbanization of their base an immediate, concrete possibility. Much debated but likely to be approved, a plan has been unveiled to transform the lower part of the river valley, remaining wetlands, stands of oleander, and orange groves included, into a huge new, high-price residential area, complete with pools, leisure parks, and golf courses.

These projects have roused the ire of many locals and new residents alike. One type of objection is expressed in terms of the aesthetic value represented by certain landscape features for the tourism industry. Since tourism is the most important industry in the area, anything affecting this sector has determining significance for local employment and welfare.[3] Another type of objection relates to the quality of life of the people who already live in the area. Higher population density brought about through the urbanization of the lower river valley without correspondingly greater services, such as new networks of roads, would worsen living conditions. If the infrastructure were increased to the new levels required, this would mean more roads, more parking lots, more suburban malls, more water damming projects. These changes, however, would in turn affect the aesthetic value of the area and once more alter the perceived quality of life of long-time residents. Still another type of objection gets raised by people who argue, for example, that the relatively wild condition of the river valley and the *posidonia* fields represents ecological values that are increasingly rare in the area.[4]

Proconservation Arguments and the Recognition of Autonomy

These types of relatively popular objections to interventions in natural environments can provide us with a baseline for reflection on the types of appeals that might be made to protect nature. The arguments contained in the objections appeal to long-term self-interest, aesthetic considerations, and nature's intrinsic values, respectively. The kinds of arguments offered in the context discussed above more or less mirror those offered in other circumstances in which natural features are endangered by human activity and may be supplemented by arguments based on traditional-societal and religious values in those cases in which they are congruent with, or promoters of, preservation or conservation of nature.

As noted by various environmental philosophers, such as David W. Ehrenfeld and Mark Sagoff,[5] proconservation arguments appealing to prudential considerations are always in danger of being overturned when conditions change. For instance, the long-term interest in maintaining certain landscape features for the tourism industry may conceivably be trumped, at some future point in time, when short-term interests of the same industry cannot easily be satisfied because of scarcity of land on which to build. Similarly, once the new infrastructure is built, it is likely that, over time, the inhabitants of the area would adjust their aesthetic valuations in such a way that they would come to consider the added roads, parking lots, malls, and dams as relatively unobtrusive to their appreciation of the overall aesthetic qualities present in the area.

Traditional-societal values also are subject to change for various reasons. For example, once a society undergoes cognitive colonization through global mass media, its values change. Because they are highly institutionalized, religious values change more slowly, but in today's world of diverse secularizing trends, often impelled by the incorporation of traditionally living populations into global mass markets, new generations may simply stop adhering to those values. These reflections lead us to consider proconservation arguments based on the intrinsic value of natural features.

Much progress has been made in clarifying how the logic of arguments appealing to the intrinsic worth of some entity is (or is claimed) to work.[6] It is quite clear that appeals to the value of nonhuman nature in itself need not be anthropo*centric*, since, *ex hypothesi*, the valuing activity is centered on *nature* and not on human beings. Holmes Rolston III, among others, argues moreover that the intrinsic value of nature should not be considered anthropo*genic*, either, because this value arises in nature's *own* teleology at organismic, species, and ecosystem levels. In other words, for Rolston, the intrinsic value is an emergent property that arises from nature's implicit goal-orientedness at each of those levels. Rolston's account constitutes a fascinating proposal, which is generating significant discussion. Notably, the issue revolves around the objectivity of values in the sense that something simply *has* a certain value just as it may have a certain other quality, such as a color.

The metaphysics of intrinsic values is much debated, of course, for the same reason that attributions of color are problematic once we assume that those attributions depend on a (certain kind of) perceiver. Without attempting to settle this question here, we may nonetheless speak about the *activity* of valuing some entity a certain way. Roughly, it seems that we may value an entity in our world in three different ways: for itself, for its instrumental utility to some individual or group of individuals, or for its value to some being whom (or which) we value for itself. Since we have a fairly clear idea of what it is to value some entity instrumentally as well as what it is to value some entity for the sake of some other being whom (or which) we value, this classification requires that we clarify what it means to say that we value some entity *for itself*. My suggestion is that valuing a being for itself requires recognition of it *as autonomous*.[7]

Autonomy literally means setting oneself one's own law and hence implies the capacity for ruling one's self. Less pompously, recognizing some being as autonomous means realizing both that this entity can maintain its organization (at least for a time) in the presence of diverse external forces and, consequently, that it may exert a systematic force on its environment, at least passively, insofar as it seeks to maintain its integrity. In other words, this means that one perceives the entity as sufficiently unified to be both source and target of organizing forces and dynamically structured in such a manner so as to maintain its unity (at least for a time).[8] This recognition of unity and of capacity to maintain itself unified is what may lead to the consideration of such a being as similar to ourselves *as selves* and, precisely because of its analogical standing to our selves, may generate our *moral* consideration.

Of course, recognizing some entity as autonomous does not mean that one is supposing it to be fully independent of all other beings, since human beings (human selves) can be recognized as autonomous even if highly dependent on many other people, culture, genetic makeup, and so on. In fact, self-maintenance may only be possible for a being while *in relation* with certain other beings or in determinate circumstances.[9] Neither am I suggesting that the recognition of autonomy should rule out all other considerations or values. We may have reasons for excluding from our moral community certain entities even if deemed autonomous, such as bacteria harmful to human beings or dangerous psychopaths. Since we are speaking here of the *activity* of valuing (notwithstanding the importance of the metaphysics of values), I limit myself to the claim that, when we *do* hold something as valuable for itself, and consequently as a candidate for moral consideration, we are doing it, among other things, *in virtue of our recognition of its autonomy*. In other words, what I propose is that, when we say that we value some being for itself, we are saying that we minimally recognize it as counting in a manner similar to ourselves, namely *for* its *self*, thereby implying that there may be legitimate, morally relevant limits to our own acting.[10] The precise manner in which such a being does morally count may still remain open at that point.

Arguments based on the intrinsic value of nature have a significant advantage over other types of arguments, such as those listed earlier, since they are not primarily dependent on a feature properly belonging to the one making the arguments. As noted, arguments based on the valuer's self-interest or aesthetic perspectives may become inoperative once conditions sufficiently change. Arguments based on traditional-societal and religious values similarly are dependent on historical developments of society and religious institutions that have nothing to do with nature as such. Arguments based on the intrinsic value of nature, in contrast, appeal to features that belong to the thing itself and as such, in principle, should enjoy stability as long as their object does.

As noted, if, when speaking of the intrinsic value of nature, we limit ourselves to that which *we* are *valuing* in nature for itself, then it seems that we are reintroducing a dependency on the one making the arguments. Insofar as arguments based on the intrinsic value of nature rely on the recognition of its autonomy, those arguments are dependent on the capacity of individuals to recognize it. I think that this dependency on a subject must be granted but does not make the arguments any less legitimate. The dependency in question goes little beyond the necessity, in the presentation of arguments, for there to be someone with a point of view. The objectivity of such arguments is dependent on their openness to intersubjective exploration. This is a topic that cannot be developed in depth here, but suffice it to say that for someone to conceive and present herself as having a self, she must already be capable of conceiving and perceiving another self differentiated from her. Recognizing the autonomy of nature requires applying this minimal personal capacity of openness to *other* selves to the realm of nature. Consequently, since arguments based on intrinsic value are based on the constitutive capacity of selves to function as such, they should be relatively protected from the vagaries to which arguments based on self-interest, aesthetics, and traditional-societal and religious value systems are subject.

Of course, arguments based on the re-*cognition* of nature's autonomy can be more or less sophisticated, since they are dependent on the knowledge acquired about nature. Hence individuals searching for proconservation arguments focused on preserving a particular species, ecosystem, or process do well to focus on the particular ways in which that species, ecosystem, or process manifests its unitary character and its capacities for maintenance of unity.

So, if valuing an entity for itself means recognizing its autonomy, then it becomes imperative to develop ways of knowing adequate to this task. Natural science has an important role here since it likely is the most sophisticated way of knowing the natural environment available to us. When put in the service of industry, though, natural science tends merely to focus on the determination of nature's goods as resources for production and sinks for waste. (This is the case, for instance, even when the value of the integrity of the rainforest is praised, since its integrity is often only valued, for example, because of its function as a repository of future pharmaceutical information.) This situation implies that we need to be clear on which ways of knowing nature do not instrumentalize it and thereby obscure its self-maintaining character. This primarily is a task for environmental education.

Concerning the moral import of recognizing the autonomy of nature, much depends on what precisely is meant by the term "autonomy." In agreement with Rolston, John Rodman, for example, proposes that "one ought not to treat with disrespect or use as a mere means anything with a *telos* or end of its own—anything that is autonomous in the basic sense of having a capacity for internal self-direction and self-regulation."[11] Eric Katz goes beyond the idea of internal teleology by, more broadly, arguing that "direct moral consideration and respect for the evolutionary processes of nature" requires that "we must respect Nature as an ongoing *subject* of a history, a life-process, a developmental system." (emphasis added).[12] Kate Soper describes the autonomy of nature as independence from human purposes, asserting that, among those who appeal to the intrinsic worth of nature in their proconservation arguments, "there is essential agreement that nature has value independently of human purposes or perceptions, that this has to do with its autonomy of those purposes, and that this provides compelling reason why human beings should revere it and as far as possible leave it be."[13]

Even William Cronon, and some of the writers assembled in his *Uncommon Ground* to make the case that nature is a cultural construct, concedes the moral importance of recognizing the autonomy of nature. Interestingly, despite arguing that the defense of wilderness is "getting back to the wrong nature," Cronon identifies autonomy with wilderness and points out its moral significance. He says that "if we acknowledge the autonomy and otherness of the things and creatures around us—an autonomy our culture has taught us to label with the word 'wild'—then we will at least think carefully about the uses to which we put them, and even ask if we should use them at all."[14] Carolyn Merchant, in the same volume, takes nature's autonomy to consist of relative independence from humanity. She proposes a "partnership ethic" between human beings and nonhuman nature, such that human beings, "as the bearers of ethics, would acknowledge nonhuman nature as an autonomous actor that cannot be predicted or controlled except in very limited domains."[15]

Famously, Immanuel Kant spoke of autonomy in more specific ways, interpreting the term as applicable only if a being were capable of participating in the Kingdom of Ends and effectively were capable of setting out a form of life, which implies a certain morality, for her- or himself. As noted, I interpret the recognition of autonomy in an entity as minimally indicating unity and the maintenance of unity, the possession of which, I suggest, invites (but does not necessitate) our moral consideration based on a recognition of an important similarity (having a kind of self) with ourselves. I suppose that moral consideration of nature follows from the recognition of its autonomy (as sketched here) *in combination with* a generalization from existing commitments to moral consideration of human selves. It remains a matter of great urgency to determine what the moral consideration of nature, once we recognize its autonomy, should mean in practice.[16] That is one of the main topics of this book.

The Papers

This book is divided into three parts. In order to discuss the autonomy of nature meaningfully, it is necessary to be clear on two basic points. First, does it make sense to speak of nature at this juncture in the discussion carried on in environmental philosophy and environmental studies? And, second, if it does make sense to speak of nature, is it reasonable to speak of nature *as autonomous*? These are the issues discussed in the first part of the book.

The second part of the book concerns itself with the question of what it might mean to respect the autonomy of nature, given that human beings have needs and interests that cannot be satisfied without certain interventions in nature. The first issue discussed in that section of the book is whether we need to have a clear idea of the *nature* of nature in order to know what it means to respect its autonomy. The second issue concerns the question whether *any* kind of influence on nature may be seen as domination. The third issue addressed is the question whether human beings can be considered as *participants* in ecosystems in such a way that their activities may be seen to contribute to nature's autonomy.

The final part of the book continues with the analysis of the human contribution to the autonomy of nature in terms of management and restoration practices. The first issue raised is whether management practices, such as are applied in nature conservation, actually promote the autonomy of nature or they rather constitute ways of turning nature into a client for a management-infatuated human establishment. The next issue discussed concerns the question whether apparently well-intentioned restoration efforts to free nature from supposed problems (such as exotics) often are mere expressions of the human desire to *control* nature, thereby, again, subtly thwarting nature's autonomy. The third issue considered concerns the possibility that, in those cases in which ecological restoration efforts clearly do not constitute domination of nature (even if they cannot be said to re-create nature), restoration activities may be able to engender a relationship with nature that can promote the autonomy of nature in the long term. The final issue addressed is whether, despite the difficulty involved in returning "naturalness" to an area, ecological restoration may help return autonomy to nature by contributing to its wildness and freedom. The book ends with a review of the discussion of what autonomy does and should mean in the context of human transformation, and restoration, of nonhuman nature.

I proceed, next, with more detailed discussions of each of the papers of the volume.

Part I. Nature and Autonomy of Nature: Are They Real?

Recently, a number of papers and books have been published that draw attention to the problematic history of the use of the term "nature" and to its multiplicity of meanings. It is pointed out, for example, that nature often has been appealed to in order to justify, without further argument, the sociopolitical status quo. It has also been argued that the use of this term has supported problematic forms of dualism, which have led to the identification with nature of those parts of society considered subordinate, especially women and slaves. The multiplicity of meanings of the term,[17] moreover, in combination with a postmodernist awareness of the social mechanisms by which our experienced world becomes real for a person, has led to the thesis of the "construction of nature," which itself at times has led to a certain skepticism regarding the reality of nature.

William Chaloupka and R. McGreggor Cawley, for instance, argue that what is contested in debates concerning the protection of nature is rooted in other, social and political, contexts. They are concerned with the idea that, without proper analysis of the discursive practices in which talk of nature is embedded, even proconservationist researchers end up as contributors to "a social theory" that seeks to render nature, both human and nonhuman, "fully susceptible to control."[18] Jane Bennett and Chaloupka continue this theme by claiming that, as a result of this situation, we supposedly come to overlook that "nature, like everything we talk about, is first and foremost an artifact of language."[19] William Cronon, as well as the contributors to his much-discussed *Uncommon Ground,* paints a similar picture. For example, in his introduction, Cronon generates a "guided tour" featuring nine different "versions of nature." Among others, he lists "*nature as moral imperative,*" "*nature as Eden,*" "*nature as commodity,*" "*nature as demonic other, nature as avenging angel,* [and] *nature as the return of the repressed,*"[20] intending to alert us to the idea that all these "versions of nature" are "cultural constructions that reflect human judgments, human values, human choices."[21]

Cronon and others, such as J. Baird Callicott, have wondered aloud, moreover, whether we should continue speaking in terms of "wilderness," given the cultural baggage that this term carries. In Cronon's analysis, in the United States, the notion of *wilderness* incites people to think in terms of the sublime and the last frontier. These ideas in turn lead to attempts to protect faraway, presumably "pristine," natural places, while neglecting nature as it is found close to home.[22] From Callicott's point of view, the notion of wilderness supposedly enshrines a problematic nature-culture dualism, is ethnocentric, and denies change, and its referent generally cannot be found since most so-called wilderness areas have long been inhabited by indigenous people. Protecting areas as "wilderness" furthermore may lead to dilemmas regarding whether recreational use of such spaces is compatible with the protection of their natural values. Callicott asks, for instance, whether an area should be open to hikers and river rafters even if such visitations endanger local biodiversity. As an alternative, he advocates forgoing the use of the term "wilderness" and the corresponding concept and redesignating relatively untrammeled areas as "biodiversity reserves."[23]

As may be seen from this brief survey, the objections to the use of the terms "nature" and "wilderness" (and to environmental analysis through the associated concepts) appeal to a variety of arguments: epistemological, ethical, and practical. Consequently, this volume begins by addressing the question whether it still makes sense at this juncture to work with such terms.

In the first contribution to part 1, Val Plumwood systematically argues that, despite the difficulties adduced by diverse authors, the urgency represented by the current degradation of the natural environment fully justifies the use of the term "nature" and corresponding appeal to the *notion* of nature, if appropriate care is taken. She claims that the nature-skeptical arguments, which take as their point of departure the claim that nature is a mere social construct, are subject to substantial critiques and that, despite the highly problematic ways in which the notion of nature has been utilized in the past, it is possible to conceive of a progressive or liberatory naturalism. In fact, Plumwood makes a powerful case for the view that, while postmodernist writers have pointed out the common failure to account for the *human* contribution—cultural and productive—to both the shaping of nature and its concept, these same thinkers have frequently overlooked the flipside contribution of *natural* creativity and *nature*'s labor to human creations and productivity.

On independent grounds from those presented by Plumwood, Keekok Lee, in the next chapter, also argues against the view that skepticism regarding the reality of nature necessarily follows from social constructionist theses. Her main point in this regard is that we have strong reasons to believe that something corresponding to the notion "nature as the nonhuman" long antedated human beings, human consciousness, and human conceptualizations. Lee proposes furthermore that nature greatly exceeding the anthropogenically affected terrestrial surface exists independently of humanity even today, thereby proving false the thesis of "the end of nature."

By introducing distinctions among various forms of teleology, Lee furthermore makes the case for the idea that natural entities, both biotic and abiotic, if independent of human intentionality and agency, may be attributed autonomy, which she calls "existence *for itself*." Lee leaves open the question of what moral significance this form of autonomy may entail, though she points out that, despite the "intrinsic/immanent teleology" supposed in such cases, we should resist assimilating natural entities to full personhood in the Kantian sense.

Part II. Autonomous Nature and Human Interests: Are They Compatible?

The diverse authors assembled in this volume have a variety of ways of understanding the autonomy of nature but, generally speaking, agree on the idea that it is constituted by a sort of independence, self-rule, unimpeded development on its own, and similar notions. An underlying assumption, which (even if not usually made explicit) seems quite reasonable, is that, if some being as such is ascribed positive value, this is prima facie grounds for respecting it. Now, given that the term "autonomy," as noted, implies the idea of independence, self-rule, unimpeded development on its own, and similar notions, and given that these are accepted to be positively valuable, it is generally taken for granted that beings with autonomy, prima facie, are worthy of moral respect.

This conclusion seems to be correct even if there may be countervailing conditions that might narrow down the cases in which moral respect is appropriate. So, if we have other important values that might be put in jeopardy by respecting the free development of a certain being, then it becomes a matter of judgment whether its autonomy is reason enough to grant it respect after all. In the human case, we may note that it would be questionable whether an individual identified as a thug should be worthy of moral respect (at least while acting as a thug), especially if we identify this individual as acting fully autonomously. Similarly, even if we have reason to attribute autonomy to a certain species of bacterium, rodent, or fungus, we may still have good reasons (stemming from other values, such as concern for human health) not to promote its development, at least within close human proximity. But, according to the reasoning proposed above, it makes sense that, once it is granted that nature or some part of it justifiably is ascribed autonomy, we at least grant it moral considerability.[24]

More than argue for the cogency of respect for the autonomy of nature, part 2 concerns itself with the question of *whether respect for the autonomy of nature may be compatible with certain human uses of nature* or whether a strict hands-off policy is called for. This is a crucial issue since environmental ethics needs to confront the reality of humanity's needs and desires for using and transforming at least some parts of nature for its own purposes.

Eric Katz's contribution connects with Lee's paper by reopening the question concerning what the autonomy of nature may be. As noted earlier, Katz has argued elsewhere that we should consider nature as a "subject of a history, a life-process, a developmental system."[25] As such, for Katz, nature's autonomy does not consist just in independence from human intentionality and agency (Lee's definition) but consists in the opposite of domination, and respect for the autonomy of nature hence is respect for a subject, thereby requiring willingess not to dominate it. Katz's paper also ties in with Plumwood's contribution, insofar as he makes reference to the difficulties of determining "*the nature* of nature" but argues that we may still know what it is to *respect* nature. Katz claims that we need not be concerned if we lack a metaphysics of nature since we do know, from our interactions with human beings, what it is to dominate and what it is to liberate a being. The "liberation of nature" hence is a matter of voluntarily withdrawing from interfering in its development. Practically, this means to "let things be, [to] leave nature alone."

Interestingly, Katz invites us to test his intuitions on a case that he describes as a "hybrid environment": an island (Fire Island, New York) that has some wilderness and also many humanly transformed areas. The problem that needs to be faced there is whether, in order to counteract beach erosion, we ought to apply sand replenishment or some more drastic form of beach stabilization (for example, the building of "permanent structures such as rock jetties and sea walls"). Katz concludes that, even if a pure "hands-off" policy cannot be implemented, "we ought to lean toward leaving nature alone," which, despite his disagreement with restoration practices, for him means that sand replenishment ought to be pursued. I will return to Katz's views on nature restoration in my discussion of part 3; now, I merely note that Katz is here offering us a particular perspective on the ethics of human cohabitation with nature: even if a certain part of the environment cannot be called autonomous in the strict sense promoted by him (because it is not entirely the subject of its own history or development), people ought to leave it alone as much as possible for the sake of the wild, autonomous parts, which somehow may benefit from this treatment.

Ned Hettinger faces the problem of human cohabitation with, and use of, nature head on. He elaborates a concept of "nature's autonomy in relationship with humanity." Hettinger thinks that respect for nature's autonomy leaves a legitimate place for the intercourse of human beings with nature. Echoing Immanuel Kant's account of ethics, he proposes that we need to explain "how humans can use nature as a means, without necessarily using it as a *mere* means." Reminiscent of Carolyn Merchant's "partnership ethic" mentioned above, Hettinger's idea is that we develop a "partnership relationship between humans and nature," which might be mutually beneficial to both. While Hettinger agrees that "the most fundamental sense of nature's autonomy is freedom from human domination and control," interestingly, "respecting the autonomy of others does not mean avoiding interaction with or influence on them."

Hettinger's perspective is subtle: domination of nature is a matter of "supreme determining or guiding influence." As long as the influence is lesser or has a counterweight in nature's influence on human beings, the autonomy of nature is maintained. Consequently, in Hettinger's account, "rural lands and domesticated animals and plants," which for Katz and Lee are rather artifactual, "can be just as autonomous as wild nature." Presumably, if faced with the question of how to address the problem of beach erosion on Fire Island set out by Katz, Hettinger would say that, given the powerful "influence" exerted by nature on the island's human inhabitants and their structures through wave and tide action, this environment may still be considered *autonomous*, and remedial action to stem erosion would *not* undermine nature's autonomy!

William Throop and Beth Vickers continue with Hettinger's particular concern for respecting autonomy on rural lands. Leaning on Aldo Leopold's idea that human beings should recognize themselves as "plain members of the biotic community" and hence may be active participants in the shaping of those communities, they are on a search for a "positive image of how we should pursue agriculture and forestry." In tune with Hettinger's focus on the degree of human influence on nature, the clue for Throop and Vickers is to "distinguish between influences that are compatible with autonomy and those that constitute domination."

Throop and Vickers ground their idea of autonomy in its etymology, interpreted as "self-rule" (*autos* and *nomos*). While in the human case autonomy means the capacity of acting according to one's *chosen* principles, in the case of beings incapable of forming intentions autonomy translates as the capacity to behave according to "principles that characterize the nature of such beings." Consequently, one's actions exhibit *respect* for the autonomy of such a nonhuman being if "they are compatible with that entity's behaving in accord with its nature."

What does this mean for respecting the autonomy of rural landscapes? First, Throop and Vickers make a case for considering ecosystems as capable of autonomy. Second, they take rural landscapes to be ecosystems that partially owe their specific natures to human interventions over historical periods. Human communities, in fact, are taken to be essential parts of these landscapes since the activities of human beings, insofar as they are sustainable, make characteristic contributions to the functioning of such areas. Hence respecting the autonomy of ecosystems constituted by agricultural areas means not dramatically altering their existing nature. Finally, armed with this notion, Throop and Vickers draw consequences for the type of agriculture (industrial versus community-focused) that does or does not respect the autonomy of nature in rural landscapes.

Part III. Management, Restoration, and the Autonomy of Nature: A Paradox?

Part 1 of the book considers whether it makes sense to speak of nature and whether it is reasonable to ascribe autonomy to nature. The papers of part 2 address what respect for the autonomy of nature might entail, given that human beings have definite needs (such as for food, shelter, cultural goods) that can only be satisfied through intervention in at least some parts of nature. The authors featured in part 3 are concerned with the types of interventions in nature that might be supposed to be carried out to *benefit nature*. These interventions are considered under two general, and often overlapping, perspectives: management and restoration.

Since the 1970s, for a number of reasons, the natural environment has become a topic for people from all over the world, especially those living in industrialized countries.[26] Industrial development has generated ever-increasing kinds and quantities of contaminants, some of severe toxicity, that work together synergistically, creating significant dangers to human health. Urban growth and middle-class lifestyles have combined with industrial agriculture, large-scale forestry, mining, and tourism endeavors to swallow up increasing portions of the remaining, quickly shrinking wilderness areas on the planet. Along with wilderness areas, species are disappearing at a rapidly accelerating rate as a result of direct and indirect impacts of human activity.

Concern for these kinds of issues has generated a number of initiatives to manage more or less affected environments in order to avoid further dangers to human health and losses of species and wilderness. Ministries have been created with the task of testing new chemicals, managing toxic sites, creating and managing natural parks, and channeling human activities in relatively wild sites so as to minimize further damage. Certain environmental nongovernmental organizations (ENGOs), such as land conservancy groups, seek to manage natural areas for the preservation of their natural values. All these endeavors seem rather congruent with respecting the autonomy of nature. Dean Bavington, however, has offered some striking considerations that suggest that, at least under certain conditions, such management efforts may *contribute* to the problems that, on the face of it, they are supposed to resolve.

Bavington's analysis is based on a concept derived from critics of health services who have argued that certain types of medical examinations and treatments in fact *generate* the need for medical assistance, or at least perpetuate it. These so-called iatrogenic effects have been detected in a variety of service interventions. In Bavington's account, environmental management efforts, as are found in environmental assessment protocols, park management regimes, or responses to natural resource crises, often (intentionally or unintentionally) *perpetuate* a need for such services, making a genuine resolution of the problems in question unlikely. These kinds of efforts at managing nature also have other problematic aspects. Insofar as management efforts generally are carried out by experts and are coordinated by individuals (belonging to some government bureaucracy, consulting company, or ENGO) who are *not* those directly affected by the problem to be resolved, those citizens who *are* directly affected thereby are turned into mere passive recipients of aid and, in effect, into clients. Bavington proposes that, in this way, management regimes applied to natural areas may undermine the autonomy of social actors.

Most important in the context of the topic of this book, besides being problematic for focusing on the satisfaction of some factional human interests, such management regimes generally treat nature and its processes instrumentally. In Bavington's analysis, the management approach is given impetus by various misguided conceptions and attitudes prevalent in contemporary society. The functional approach, for example, considers species and ecosystems merely as components in systems, the functioning of which this approach seeks to optimize, with the consequence that, under some circumstances, certain species may be sacrificed if others can fulfill their function just as well! The bottom line is that, even if apparently beneficent, according to Bavington, the application of the managerial approach to nature in practice usually neglects to consider nature as autonomous. So, as a result of the push to implement management regimes in all spheres of life, both nature and human beings become instrumentalized as clients.

Viewed in relation to the papers previously discussed, Bavington is, in effect, pointing out that respect for the autonomy of nature goes *beyond* avoiding domination, or limiting degrees of influence, or avoiding dramatic changes to existing ecosystems (humanly altered or not), as Katz, Hettinger, and Throop and Vickers, respectively, have argued. Bavington's account implies that there is something fundamentally wrong with the way human beings treat nature if their practices, although intended to benefit nature, fail to take into account the kind of relationship between human beings and nature that they create or exemplify. Managerial approaches, however efficient, turn citizens and nature into objects, while respect for their respective forms of autonomy requires that they be treated as subjects.

John Sandlos continues with the question whether the managerial approach to nature is truly agreeable with the autonomy of nature. Sandlos considers a particular example of a management effort in North American wetlands concerning a plant called purple loosestrife, which diverse experts and management bodies consider a weed because it supposedly displaces native species. Sandlos finds a number of problems in the handling of this situation, beginning with the assumptions that natural science truly is objective, that natural science can distinguish between species that belong and species that constitute pollution in an area, and that natural science is free from social influences in its categories and motivations.

Ultimately, Sandlos takes issue with the supposition that such natural science–guided management efforts to restore nature to a pure condition, which it presumably had before so-called invasive species arrived, may legitimately be seen as liberation of nature. Sandlos concludes that the purpose of the "war" against purple loosestrife is "not the preservation of a wetland community but the maintenance of a social and economic pattern that demands dominance of human interests and influence in the natural landscape." Such attempts to free nature of "the depredations of a noxious weed" are really "part of the larger mythic battle for control over the universe" carried out by humanity. By implication, weed control measures, such as those carried out against purple loosestrife, may be rather antithetic to respect for the autonomy of nature.

Since there are large-scale incursions into wild nature, carried out, for example, in the context of unsustainable forestry or mining practices, management of relatively untrammeled natural areas increasingly is being complemented by restoration efforts. Restoration of creek beds, wetlands, or forests has been welcomed by some as a form of morally required mitigation of the harm done to natural areas by human activities. Others have offered a number of critiques of restoration, for example, on the suspicion that such efforts are intended to justify environmental damage a posteriori or that they are a kind of deception, since restoration by humans does not have nature as its origin. Most relevant from the perspective of this book is the critique, offered by Eric Katz, that nature restoration is *not* a way to return autonomy to nature since it presumably involves *domination*.[27]

The claim seems to receive support from the observation that restoration involves intentional manipulation of parts of the natural world, which, in the end, supposedly can only result in the production of artifacts. Andrew Light takes this point of view to task. Light is ready to put aside the question whether human beings can go beyond producing artifacts but insists that benevolent restorations (i.e., restorations carried out *pro bono natura*) need not entail domination. Similarly to Bavington and Sandlos, he takes note of the fact that human interaction with nature generates a certain kind of relationship. While the kind of nature management to which Bavington and Sandlos allude instrumentalizes nature, issuing in domination even if presumably intended for its benefit, Light proposes that benevolent restorations can generate a "positive normative value."

Light's claim is based on the proposal that the relationship with nature *in itself*, not just "the way we interact . . . in the relationship," has value. One of the most important tasks for environmentally concerned citizens hence is to find ways to develop a genuine, caring relationship with nature in the first place. Light says that "we need some kind of material bridge . . . to see that relationship come about." Since in benevolent restorations we are not only "bound *by* nature" but also "bound *to* nature," such activities are excellent candidates for the generation of an appropriate relationship.[28] According to Light's analysis, such activities may bring about the development of a "culture of nature," that is, a kind of culture that recognizes the autonomy of nature since it presupposes very close attention to the functioning of natural processes.

Light points out that, complementary to the relationship that restoration may help develop with nature in general, restoration activities may further the goal of supporting the autonomy of nature in a variety of other ways. It may, for example, bring about the creation of areas that, through their resemblance with pristine nature, can increase "bonds of care that people have with nonrestored nature," help in the generation of better understanding of the "actual consequences of human domination of nature," or facilitate processes that "allow nature to engage in its own autonomous restitution."

Light's analysis interestingly resonates with Hettinger's and Throop and Vickers's papers, which argue that human influence on nature need not be equated with domination but may even lead to its flourishing as autonomous. Moreover, although Light's paper directly seeks to counter Katz's view of restoration as domination, ultimately the two authors seem to agree that, in the face of environments heavily affected by human activities ("hybrid environments" in Katz's terms), there are still ways of being more rather than less respectful of nature's autonomy. For instance, Light would likely agree with Katz that sand replenishment of the eroded beaches of Fire Island is a better option than the construction of rock jetties and sea walls. The difference between Light's and Katz's perspectives mostly seems to be a matter of emphasis, since for Light such restorative efforts have the added value of facilitating not only nature's self-recuperation but also the generation of relationships, and associated attitudinal changes, that may serve the preservation of nature's autonomy in the long term.

The final paper in part 3, by Mark Woods, is concerned with questions concerning both management and restoration in the context of respect for the autonomy of nature. Woods examines to what degree management of a wilderness reserve area, such as the Everglades National Park, is compatible with properties such as naturalness, wildness, and freedom, where the apparently required management regime includes ecological restoration. Like Light, Woods is aware of the difficulty (argued by Katz) entailed by the supposition that human beings can return naturalness to an area without turning the area into an artifact, since naturalness is "a relational property that refers to a causal history characterized by nonintentional, ateleological physical, chemical, biological, and evolutionary forces." According to Woods, we may come to realize, however, that management of areas whose natural values have already been negatively impacted by human beings can benefit from ecological restoration if we distinguish the attempt to return *naturalness* from the attempt to return *wildness* and *freedom*.

In contrast to naturalness, which depends on an area's specific causal history, wildness, rather, describes a form of autonomy belonging to a "more-than-human world" in which events "occur largely because of [that world's] own internal self-expression that is independent of civilized forces." Freedom in this context moreover "connotes having a broad range of opportunities for self-expression that is unhindered from confining, external forces." Given these distinctions, Woods notes that, even in areas heavily impacted by human activity, certain animals and plants may come to express wildness, or autonomous development, and exhibit a certain degree of freedom, if their development mostly is a result of their own internal properties and not the result of human direction and if they are not confined by humanly created barriers and impediments, respectively.

In this way, Woods intends to respond to Katz's critique of restoration, according to which such practices (in Woods's words) "trammel . . . *currently* existing wildness and freedom" (emphasis added). Woods's response is that, where restoration is justified, such (minimal or nearly nonexisting) wildness and freedom is probably worth sacrificing on behalf of the development of "*future*-oriented wildness and freedom" (emphasis added). Woods's paper complements Hettinger's and Throop and Vickers's essays insofar as he defends the congruence of certain kinds of influence with the respect for autonomy. He also sharpens the points made by Bavington and Sandlos, who argue against certain sorts of management and restorative practices, by making clear that the kind of management and control that they object to would undermine autonomy as represented by wildness and freedom. Finally, Woods's paper agrees with Light's arguments to the effect that restoration need not imply domination even if it involves manipulation of natural beings, processes, and spaces, since such manipulation may issue in "diachronic," or future, expressions of autonomy in the wildness and freedom of natural beings.

Even if up to this point the papers, in their distinct perspectives, sometimes appear to be in conflict with each other, they nonetheless all emphasize Plumwood's point that nature is a legitimate category for the exploration of the relationship of human beings with the nonartifactual, nonhuman world, and they all confirm Lee's claim that nature (at least under some conditions) may justifiably be conceived as autonomous. The book ends with a critical discussion/assessment of the preceding papers by invited commentator William Jordan III, who takes to task the notions of nature and autonomy when they are used as easy ways of escaping the ineluctable ambiguity of human existence, which is both natural and other than natural. This does not mean, however, that Jordan has no use for the notion of autonomy. Autonomy, seen as relative independence of natural entities founded in evolutionary and ecosystemic relations, is central to his account of restoration, which, in his view, consists in returning as much wildness to natural processes as human beings possibly can. In this way, Jordan challenges at least some of the preceding authors by suggesting that it is precisely human intervention that can return autonomy to nonhuman nature.

The papers in this volume offer a rich diversity of viewpoints on what the autonomy of nature is, how it may be protected, and how human activities may or may not interfere with it. What remains to be done is an exploration of how respect for the autonomy of nature may be translated into everyday morality, politics, and economics.[29] This is a task that, in addition to sensitivity to conceptual distinctions, will also require good will and attentiveness to the diversity of nature's expression. Ultimately, the recognition of the autonomy of nature, in both its epistemological and moral sense, calls for a new form of attending to, and caring for, nature, which we might a "culture of nature."[30]

NOTES

1. Some of the contributors, notably Keekok Lee, furthermore understand the autonomy of nature in ontological or metaphysical terms.
2. Some environmental philosophers, including Bill Devall and George Sessions, argue for a distinction between conservation and preservation. Here, I use "conservation" as an umbrella term for protection of natural beings and the processes that maintain them, be the motivation for protection human self-interest, identification with nature, care for nature's intrinsic value, or something else.
3. Sometimes these sorts of concerns have led to changes in municipal laws and regulations. It has been made illegal, for example, to replace the old core of white hilltop houses of my hometown with high-rise apartment buildings or to build on the upper part of the mountains guarding our bay.
4. The problems my hometown, Altea, faces are not unique. Many towns along the Mediterranean shores are in similar situations.
5. See, for example, David W. Ehrenfeld, "The Conservation of Non-Resources," *American Scientist* 64 (November–December 1976): 648–56; and Mark Sagoff, *The Economy of the Earth* (New York: Cambridge University Press, 1988), especially chap. 6.
6. Especially see J. O'Neill, "The Varieties of Intrinsic Value," *The Monist* 75 (1992): 119–37.
7. My discussion concerning autonomy here is not intended as a critique of the diverse conceptualizations of this idea offered further on in the volume but rather as a kind of a *prolegomenon*.
8. I thank Mark Woods for helping me think more clearly through these points. He suggests that the autonomy of nature "contrasts with obedience: wild things are autonomous because they have not changed to adopt the imposed will of another. We can also think of autonomy in terms of authenticity: being self-expressing, self-actualizing, or self-realizing" (Mark Woods, letter, June 2003).
9. It has been argued that the concepts of autonomy and interdependence of selves need not be taken as antagonistic, as noted in Virginia Held, "Feminist Transformations of Moral Theory," in Steven M. Cahn and Peter Markie, eds., *Ethics: History, Theory and Contemporary Issues* (New York: Oxford University Press, 1998), pp. 682–99, esp. p. 695, making reference to Jennifer Nedelsky, "Reconceiving Autonomy: Sources, Thoughts and Possibilities," *Yale Journal of Law and Feminism* 1 (spring 1989): 7–36, esp. p. 9.
10. In this context, also see Val Plumwood, "Nature, Self, and Gender: Feminism, Environmental Philosophy, and the Critique of Rationalism," in Christine Pierce and Donald VanDeVeer, eds., *People, Penguins, and Plastic Trees*, 2d ed. (Belmont, Calif.: Wadsworth, 1995), pp. 197–213, on the danger of supposing that we should identify our selves with nature through an extension of our selves, possibly "failing to recognize unambiguously the distinctness and independence of the other" (p. 205).

11. John Rodman, "Ecological Sensibility," in Donald VanDeVeer and Christine Pierce, eds., *People, Penguins, and Plastic Trees* (Belmont, Calif.: Wadsworth, 1986), pp. 165–68, quotation on p. 166.
12. Eric Katz, *Nature as Subject: Human Obligation and Natural Community* (Lanham, Md.: Rowman and Littlefield, 1997), p. xvi.
13. Kate Soper, *What is Nature? Culture, Politics and the Non-Human* (Oxford: Blackwell, 1995), p. 252.
14. William Cronon, "The Trouble with Wilderness; or, Getting Back to the Wrong Nature," in William Cronon, ed., *Uncommon Ground: Rethinking the Human Place in Nature* (New York: Norton, 1996), p. 89.
15. Carolyn Merchant, "Reinventing Eden: Western Culture as a Recovery Narrative," in William Cronon, ed., *Uncommon Ground: Rethinking the Human Place in Nature* (New York: Norton, 1996), p. 158.
16. It goes without saying that the authors in this volume don't necessarily agree with my own sketch of what the autonomy of nature is. My sketch should be considered, rather, as a first move, to be contrasted and compared with the other proposals that follow.
17. Also see John Torrance, ed., *The Concept of Nature: The Herbert Spencer Lectures* (Oxford: Oxford University Press, Clarendon, 1992), for an account of the changing meaning of the term since the time of the ancient Greeks.
18. William Chaloupka and R. McGreggor Cawley, "The Great Wild Hope," in Jane Bennett and William Chaloupka, eds., *In the Nature of Things: Language, Politics, and the Environment* (Minneapolis: University of Minnesota Press, 1993), p. 5.
19. Jane Bennett and William Chaloupka, "Introduction: TV Dinners and the Organic Brunch," in Jane Bennett and William Chaloupka, eds., *In the Nature of Things: Language, Politics, and the Environment* (Minneapolis: University of Minnesota Press, 1993), p. xii. Notably, while trying to speak from the perspective of the world-as-experienced, some postmodernist writers unwittingly seem to conflate the concept "nature" with the thing: nature. Generally, however, they admit that the distinction can and must be made.
20. William Cronon, "Introduction: In Search of Nature," in William Cronon, ed., *Uncommon Ground: Rethinking the Human Place in Nature* (New York: Norton, 1996), pp. 36, 46, 48.
21. Ibid., p. 34.
22. Cronon, "The Trouble with Wilderness," pp. 69–90.
23. See, for example, J. Baird Callicott, "The Wilderness Idea Revisited: The Sustainable Development Alternative," in J. Baird Callicott and Michael P. Nelson, eds., *The Great New Wilderness Debate: An Expansive Collection of Writings Defining Wilderness from John Muir to Gary Snyder* (Athens, Ga.: University of Georgia Press, 1998); also see "My Reply," in Wayne Ouderkirk, ed., *Land, Value, Community: Callicott and Environmental Philosophy* (Albany: SUNY Press, 2000), 290–328.
24. There are more elaborate arguments for respecting autonomous nature, such as Katz's reliance on the idea that nature is a *subject*. In this view, respect is due

to nature for quasi-Kantian reasons, insofar as treating subjects as if they were *mere objects* is a straightforward moral mistake. See Eric Katz, *Nature as Subject* (Lanham, Md.: Rowman and Littlefield, 1997).

25. Katz, *Nature as Subject*, p. xvi.
26. Though not only in industrialized countries. See, for example, Thomas Heyd, "Environmental Ethics in Latin America," *Environmental Values* 13 (2004): 223–42.
27. See Katz, *Nature as Subject*, passim.
28. On the value of nature restoration for the development of the relationship between human beings and the natural environment, also see Thomas Heyd, "Nature Restoration Without Dissimulation: Learning from Japanese Gardens and Earthworks," *Essays in Philosophy* 3, no. 1 (January 2002), http://sorrel.humboldt.edu/~essays/heyd.html.
29. Regarding environmental morality, see Thomas Heyd, "The Case for Environmental Morality," *Environmental Ethics* 25 (spring 2003): 5–24. Concerning the integration of environmental morality in the workplace, see Thomas Heyd, "Environmental Ethics in the Workplace," in Robert Larmer, ed., *Ethics in the Workplace: Selected Readings in Business Ethics*, 2d ed. (Belmont, Calif.: Wadsworth, 2001).
30. I discuss the notion of a culture of nature further in "Nature, Culture, and Natural Heritage," *Environmental Ethics* (forthcoming); and in *After Nature: Encountering Nature in Hybrid Spaces* (Aldershot: Ashgate, forthcoming). I am indebted to Mark Woods for carefully looking over the essay and making a number of excellent suggestions for its improvement. In some of the notes I have directly adopted some of the language he suggested. I also would like to thank my hosts in Altea La Vella, Elke Haeberlein and Mustafa Celepoglu, for their patience during the preparation of this revised introduction.

PART I

Nature and Autonomy of Nature

ARE THEY REAL?

TWO

Toward a Progressive Naturalism

VAL PLUMWOOD

THE DEEP contemporary suspicion and skepticism about the concept and term "nature" may play some role in the contemporary indifference to the destruction and decline of the natural world around us. If the category "nature" is seen as phony, if it can appear only when suitably surrounded by sneer quotes, we are hardly likely to be inspired by appeals to nature's integrity in the case against genetic engineering or for the defense of nature in the case for stopping the current slaughter of the seas and the holocaust of animal life. The more nebulous and indeterminate such nature skepticism is, the more difficult to dispel is the general sense of unease the term "nature" seems to arouse in the modern and especially the postmodern mind. Even if that unease can be justified for some areas of usage, the danger is that it will contaminate perfectly defensible and useful, even indispensable, roles for concepts of nature, in a way that will make important conservation causes very difficult to articulate convincingly. Should we then abandon "nature" as the banner term under which we might try to resolve the ecological crisis? I suggest the answer is no. One aim of this paper is to help distinguish between more and less problematic senses of "nature" and "naturalism." I shall use the term "nature" here primarily in its dominant contemporary sense to mean the sphere of the nonhuman, although, as we shall see, that sense itself is not unproblematic.

This clarification project may help clear some of the difficulties in the way of imagining a progressive or liberatory naturalism, that is, a project that aims to draw together oppositions to oppressive forms and narratives that have made use of culture/nature or reason/nature dualism. To the extent that this dominant culture has often coded and treated the oppressed side as a form of nature,[1] a progressive politics based around a critical and partial affirmation of nature would seem appropriate, rather than a simple rejection of the coding itself (although in many cases both are required). There are various directions from which such a project of coalition or political convergence might be approached: in the hands of ecofeminist theorists, this objective has been based on the potential of the concept of nature, in its various historical guises, to support oppositions to a certain web of oppressive forms, oppositions that characteristically make up progressive politics. Ecofeminists among others have noted that many oppressed groups—including nonhumans, women, people of non-Western culture or races, those who perform manual or bodily labor, and others identified as of a lower, supposedly less rational class—have often been envisaged in Western culture as less human, as a form of nature or as closer to "nature," in opposition to reason, culture, or civilization, the latter group of concepts often carrying a strong identification with privileged groups.[2]

The unificatory advantage and potential of progressive naturalism would be based on its ability to collect oppositions to a key set of centrisms constructed around a version of the concept of reason (or a reason-based concept of culture), which includes a common cultural narrative that guides the selection and setting apart of privileged groups: that of the hyperseparation and superiority of reason to nature. As the One is to the Other, as male is to female, as modern rationalist culture is to cultures it defines as lesser, so are those identified with reason/human/civilization/culture to those defined as closer to nature. Such a project of progressive naturalism might tap deep resonances and strong historical currents of opposition and resistance in Western culture; for example, it could connect usefully with some progressive forms of the romantic movement (although the romantic connection can also serve to indicate some of the project's dangers).[3]

Difficulties for a Progressive Naturalism

A project of progressive or liberatory naturalism faces many difficulties, which I discuss below. But even if it never overcomes these sufficiently to become a popular intellectual rallying point for liberatory politics, it is necessary to go a certain distance along the road to constructing it in order to understand the contemporary politics of nature. An investigation of the old tension between progressive and naturalistic politics can help to show us when and where we may need to adopt naturalizing versus denaturalizing strategies, as I explain below. And it can further the crucial cultural project of changing the dominant narrative and its leading characters of reason and nature.

The term "progressive naturalism" itself has contradictory overtones, since the concept of progress has long been defined as the adjunct of reason, the side opposing nature in the West's dominant cultural narrative of reason progressively mastering nature that justifies maximizing rational control of the earth. But in this account the terminology is all the more useful in flagging the intention of a liberatory synthesis that disrupts the reason/nature oppositions and narratives that have framed these concepts and defined their characteristics within a colonizing relationship of human centeredness, with its attendant histories of domination and projects of subsumption or assimilation. The tension in the idea of progressive naturalism, like that in the idea that of "rational woman,"[4] points to what cannot be articulated in a way consistent with the definitions established by the dominant framework.

If naturalism aims to articulate a basically affirmatory, countercultural stance toward the natural (including the biological, corporeal, and material) world and our own human inclusion within it, the concept of progress might aim to flag naturalism's intended alliance with the liberatory politics of critical social change movements. For such an alliance to be successful, the concept of progress must be given a new role outside the dominant reason/nature narratives that have linked it with the mastery of nature. Reworking is necessary for several reasons.

Despite current temptations to cynicism about the concept of progress, I do not favor any simple abandonment of the concept as inevitably complicit in those dominant narratives; some concepts (like democracy and progress) are just too important to cede to the opposition. Reworking of progress concepts in different terms is part of a reworking of the dominant narrative or reason opposing nature and its leading concepts.[5] And the old oppositional story is more inappropriate than ever in these times: a concept of progress can no longer just be opposed to that of nature but has to be redefined to recognize nature's limits.[6]

Progressive naturalism requires that we disentangle the liberatory roles of the concept of nature from the antiliberatory ones. This can be difficult. Culture/nature dualism hinders this disentangling project in several ways; it legitimates usages in which categorizing something as natural functions to rule out any social focus, any questioning or resistance. In antiliberatory cases (which I will call conservative naturalism), the concept of nature is used to minimize the contribution of the social and thus the need for social change. Conservative naturalism seeks to naturalize oppression, invoking nature to universalize and justify, to depict as "natural" and unalterable, oppressive arrangements that are actually contingent and quite open to change. In these cases, it may be objected initially that the inclusion of the relevant group, process, or state in the category of nature in a way that excludes or demotes the role of the social or cultural is mistaken and ideologically motivated. But, given the historical web of identification of oppressed groups with nature and its implications, this inclusion in the category of nature is rarely just a matter of innocent misclassification. The invocation of nature in such contexts often serves to suggest inevitability and exclude sociocultural explanations and remedies, functioning in colonizing frameworks to naturalize inferior treatment.

Progressive naturalism may appear to be swimming against the current stream of progressive and constructivist positions that insist on full recognition of the social and cultural elements in human lives, in opposition to conservative naturalism and individualism. I will argue though that there is not only no necessary opposition between recognizing the hidden or denied social and cultural elements and recognizing the hidden or denied elements of agency in the sphere of nature but that these two kinds of recognition mostly need and assist one another. Colonizing perspectives find the category of nature serviceable both to suppress resistance and to hide certain kinds of (human and nonhuman) inputs they wish to appropriate, refusing to recognize the suppressed other's agency and creation of value and assimilating relevant cases to that of nature. Usually this is possible because, within the dominant narrative, nature's agency is itself denied, so that to be included within the category of nature is to be deprived of recognition as an agent. I discuss the logical dynamics of this important class of cases in detail in the next two sections, arguing that it does not justify the dismissal or denial of the category of nature itself. Indeed, to the extent that nonhuman species have their own forms of culture, agency, and autonomy, the exclusionary opposition between nature and culture is simply invalid and depends on an oppressively reductionist view of nonhuman animals (which may then be read back into selected human cases to oppress them also).

Taking account of these cases of the use of the concept of nature to oppress, however, goes only part of the way toward explaining why many progressives in the conventional left sense have tended to be hostile to the idea of nature and seek everywhere to distance themselves from it and to minimize its extent and importance. Progressive naturalism would insist that the conventional "realo" camp has ceded the concept and sphere of nature somewhat prematurely to the enemy. Other reasons for the traditional hostility of progressives toward nature include the tendency of the traditional left to frame its project in terms of some version of the dominant narrative of reason/nature dualism and domination. Within this narrative, movement out of the oppressed category can only be achieved by distancing from the sphere of nature and the oppressed condition of being coded within it.[7] To this we may add the influence of Marx, whose devaluation of the aspects of the human shared with other animals, of nature as a sphere of rural idiocy and cultural impoverishment, and of peasant revolution is well known.[8]

Further motives for oppressed groups to derogate and distance from nature as the nonhuman may include the repetition of the privileged gesture of exclusion directed toward a still-lower, more fully inferiorized group. This gesture of exclusion may form part of a case for inclusion in the more fully privileged categories, (e.g., fully "civilized" humans), or it may be an overreaction to the misclassification or coding as nature that tends to form part of the mechanism of oppression.[9] In some cases, it may be based on the belief that it is best to minimize the number of outsiders (e.g., nonhumans) who can join privileged insider groups in the interests of maximizing the individual welfare of insiders (humans) and their ability to improve their lot by exploiting outsiders.[10]

Postmodern tendencies to dismiss or reduce the category of nature are discussed by Kate Soper in her 1994 book *What is Nature?* and include the proclivity for reducing everything possible to culture. However, many postmodern concerns about essentialism and claims to cultural universalism point to important issues. When questions of different cultural relationships to nature are under discussion, vital questions may be begged unless we recognize that "nature" is a Western concept that is not culturally universal. Sometimes we can rephrase our concerns in more culturally pluralistic terms, for example, as questions about different cultural relationships to land or (where relevant) to (specified parts of) the nonhuman sphere. And sometimes we should switch to a less homogenized set of contrasts and hegemonic wholes.

Any project of progressive naturalism faces major problems. It must avoid a reverse centrism of substituting nature for culture or reason in a new narrative of dominance that mimics and parallels the old, failing to reimagine with sufficient daring the identities and relationships the original narrative created. This tendency to reverse centrism or domination of nature over reason has sometimes afflicted parts of the romantic movement.[11] The problems of an unreconstructed reverse affirmation as a solution to gender dualism are all too evident from the case of women, where an uncritical affirmation of "women's virtues" and of character ideals of womanliness defined in the private sphere can serve to keep women as strongly regimented and home-defined as ever. Any affirmation of nature requires the same kind and level of qualification, at the very least.

Another problem is that some of the key identifications with nature have waned as explicit doctrines at the ideological level, even if the oppression they once justified continues at the level of practice in different forms and guises. The idea that human others are less human because they are more natural or animal is in many places now rarely expressed in bald terms. The Great Chain of Being hierarchy it once theorized has become one of the background resonances of the culture rather than a conscious contemporary doctrine we often meet out in the open. This does not mean that such webs of connections are unimportant—indeed, to the extent that they are operative but not consciously so, they are more rather than less dangerous and in need of exposure—but it means that complex and often abstract argument is required to establish at the political level the common ground any affirmatory stance toward nature might provide. This potential to recover common ground can be rendered problematic, too, by other developments, for example, the degree of horizontal violence among the oppressed groups such a naturalism might hope to solidarize.

Nevertheless, exploring this potential common ground remains an important political project because of the structural similarities the dominant narrative of nature has bequeathed to the diverse forms of oppression it has justified. Nature provides the model in terms of which other denials are framed and excused, as its own prior agency is denied and overridden in the foundation of property.[12] Hegemonic conceptions of human agency are fostered in human-centered culture; these are linked to denials of dependency, which are in turn linked to the application of inappropriate strategies and forms of rationality that aim to maximize the share of the isolated self and neglect the need to promote mutual flourishing. Thus the One or Man of Property is able to assume the contribution of nature in the form of a continuing support base for production, accumulation, and renewal but to deny it in failing to recognize and allow for nature's reproduction and continuation.[13]

A sufficiently careful and well-articulated form of nature affirmation might be able to negotiate these difficulties. The chief political problem for the present, though, one I address below in the third section, is to separate with sufficient clarity and permanence the oppressive and the liberatory roles the concept of nature can play. Although there are reasons for skepticism about the project of using a reconstructed concept of nature as a political rallying point, it is an important clarificatory exercise, in attempting to meet this and other difficulties, to ask what reconstructions of the concept of nature might support such an affirmation.

Naturalism in the form of some kind of recognition and affirmation of the larger nonhuman sphere and of our own animality[14] needs sympathetic exploration as a necessary corrective to the excesses of many alternative strategies, including the postmodern and constructivist strategies of dispensing with nature, which I discuss in detail in the fourth section below, and the more traditional course of affirming the dominant side of the dualized nature/culture pair, in the form of humanism. Humanism must come to terms with an affirmation of the denied nonhuman side of the dominant human culture that is labeled nature if it is ever to find a satisfactory form for its human applications. The oppositional affirmation of, and limited focus on, humans characteristic of humanism has supported both human inequality and human self-enclosure. It has helped us to lose touch with ourselves as beings who are not only cultural but also natural, embedded in the earth and just as dependent on a healthy biosphere as other forms of life.

Humanism is multifaceted and can include potentially positive elements and aims such as human solidarity and equality. But the position has long been open to distortion and subversion in several respects: equality has been subverted by humanism's overemphasis on reason, which has allowed an elite-based rationalism to interpret its leading concepts of reason and nature to benefit a rational meritocracy, for example, through interpreting ideas of impartiality, universality, and objectivity as involving the exclusion of care, compassion, and emotionality. Second, the tendency to build concepts of human equality and solidarity on an exclusionary form of bonding defines the human in dualistic opposition to its Other, the hyperseparated contrast class of the nonhuman. Third, the doctrines of equality and justice these positions have enunciated have often been subverted by the insistence on a sharp, even emphatic, boundary to their inclusiveness.

The exclusion of nonhumans from ethical and other forms of concern is a moral boomerang, which too often returns to strike at humanity itself when supposedly lower orders of humans are assimilated to those beyond the boundary, to nature and to animals, as they have been systematically through much of Western history. All these human-supremacist features rebound against the project of human solidarity and have been mobilized against those human groups associated with the excluded nonhuman class. Thus the second and third elements of exclusion have long done battle with the first element of equality and solidarity among humans. Humanism can only resolve these tensions by moving beyond its traditional exclusive focus on, and centralization of, humans, human agency, and interests.

Distorting Dualisms and Hegemonic Constructions of Agency

"We ought to purge our environmental manifestos of the language of the 'domination of nature,'" writes Neil Smith,[15] since domination of some sort is inevitable, he thinks. Smith seems to think this is so because use of technology, use of nature, and influence on nature are all inevitable parts of human life. Yet none of these equals domination, as we can see if we consider that all these are also an inevitable part of our interactive social relations with other human beings, where we concede, or should, that relations of domination are not inevitable.[16] Stances of domination and mastery toward the nonhuman sphere seem inevitable to many people in Western culture because they are often unfamiliar with alternative cultural patterns and frameworks and such stances form such a deep part of their framework of thought. Also the actual patterns of relationship that result from domination or mastery are rarely spelled out in sufficient detail to enable us to see how to escape them. Yet, when they are, it is apparent that such relations are by no means inevitable and that understanding how domination shapes our concepts of, and relations to, nature is very important in understanding the ecological threats the culture of mastery is creating for itself.

Relationships of domination form identities of dominator and subordinated in major ways, and much of the pattern of relationship we can discern here appears to be common to both human and nonhuman cases. Relations of domination can take different forms; however, a well-established pattern for Western culture in the case of nature involves colonizing relationships justified by anthropocentrism, just as the intrahuman relationships in which Europeans colonized the lands of those they believed to be lesser were justified and supported by Eurocentrism. In both cases, relationships of dualism or binary opposition are created around the identities of the One and the Other, in this case human and the nonhuman, and the Other is treated as something to distance from and subdue. A number of diverse elements can be involved here,[17] of which I will discuss two that have a major bearing on contemporary blind spots in relation to the dependency of human culture on nature.

Hyperseparation

The first of these is *hyperseparation,* an emphatic form of separation that involves much more than just recognizing difference. Hyperseparation means defining the dominant identity emphatically against or in opposition to the subordinated identity, by exclusion of real or supposed qualities. The function of hyperseparation is to mark out the other for separate and inferior treatment through a radical form of exclusion. Thus macho identities emphatically deny continuity with women and try to minimize qualities in themselves thought of as appropriate for women. Colonizers exaggerate differences (for example, through emphasizing extreme cleanliness, "civilized" or "refined" manners, body covering, or alleged physiological differences between what are defined as separate races or classes).[18] They may ignore or deny relationship, conceiving the subordinated party as less than human. The colonized may be described as "stone age" or "primitive," as "beasts of the forest," and contrasted with the civilization and reason attributed to the colonizer.[19]

Similarly, the human "colonizer" treats nature as radically Other and humans as emphatically separated from nature and from animals. From an anthropocentric standpoint, the nonhuman sphere is a hyperseparate lower order lacking any real continuity with the human.[20] This approach tends to lay heavy stress on those features that make humans different from nature and animals rather than those they share with them as constitutive of a truly human identity. Anthropocentric culture often endorses a view of the human as outside of, and apart from, a plastic, passive, and "dead" nature that lacks agency and meaning. A strong ethical discontinuity is felt at the human species boundary, and an anthropocentric culture will tend to adopt concepts of what makes a good human being that reinforce this discontinuity by devaluing those qualities of human selves and human cultures it associates with nature and animality. Thus it associates with nature inferiorized social groups and their characteristic activities; women are historically linked to nature as reproductive bodies and through their supposedly greater emotionality; indigenous people are seen as a primitive, earlier stage of humanity. At the same time, dominant groups associate themselves with the overcoming or mastery of nature, both internal and external to the human self.

To understand why such constructions are not inevitable, it is crucial to distinguish between separation and hyperseparation.[21] Countering hyperseparation of humans from nature implies recognizing continuity and hybridity between the human and the natural. It does not require us to deny nature's otherness or separateness or to deny or submerge human distinctness from other species, for example, by the claim that humans are just part of nature. Humans *are* part of nature, in the sense that they are subject to ecological principles and have the same requirements for a healthy biosphere as other animals, but they, like all other species, also have their own distinctive species identity and relationship to nature. This need not be, however, a relationship of domination, and neither need the traits taken to characterize the human (e.g., language, rationality) be the only ones accorded esteem. To counter hyperseparation, we need a depolarizing reconception of nonhuman nature that recognizes the denied space of our hybridity, continuity, and kinship and is also able to recognize, in suitable contexts, the difference of the nonhuman in a nonhierarchical way. And we should be suspicious of hyperseparated senses of nature, since to be other (or separate, distinct) is not the same as to be purely other (or hyperseparated). A number of paradoxical and skeptical arguments trade on this ambiguity to make it seem that, because of the pervasiveness of human presence and influence, nature as the purely other, and therefore nature as such, does not exist. I discuss some of these arguments below.

Backgrounding

Another very important feature of frameworks of domination is backgrounding, a form of simultaneous reliance on but disavowal of the agency of subordinated Others. When the dominating party comes to believe that it is radically different and superior to the subordinated party, it is also likely to devalue or deny the Other's agency and its own dependency on this devalued Other, treating it as either inessential and substitutable or as the unimportant background to its foreground. Thus women's traditional tasks in house labor and child raising are treated as inessential, as the background services that make "real" work (the work of the male) and achievement possible, rather than as achievement or as work themselves. The conceptual means by which this simultaneous reliance and disavowal is accomplished is through the hegemonic construction of agency. In highly androcentric frameworks such as Aristotle's, women's reproductive agency was backgrounded as an adjunct or mere condition for real agency, which was claimed for the male reproductive role, the woman being merely "the nurse" for the male seed. Aristotle's age erased women as social and political agents, enabling Aristotle to erase women's reproductive agency in his award of the reproductive ownership of the child to the father. Aristotle saw the father as contributing the rational element of form as compared to the mother's contribution of mere matter. In this hegemonic construction, the father emerged as the only active agent in a reproductive situation, which we now conceive as normally involving joint and mutual agency.

Hegemonic Constructions of Agency

Splitting or hyperseparation and backgrounding or denial work together to produce typical hegemonic constructions of agency. This is well illustrated in the Marglins' study of dominating forms of knowledge.[22] Knowledge, which in some cultures remains integrated and fully embodied, is in Western cultures often split into a superior abstract "rational" form versus an inferiorized "practical," experiential, and embodied form, usually reflecting the different status of the different groups possessing it. The split opens the way for the dominance of abstract "rational management" over those reduced to serviceable bodies that carry out the tasks, management plans, and dictates and also allows appropriation of agency on behalf of those counted as rational managers. The dominant party can afford to forget the other, provided they continue to function in serviceable ways or are replaceable (substitutable), and, if their level of denial goes deep enough, may be inclined to do so even where the other is not replaceable.

Contemporary hegemonic constructions of agency are the other side of, and are encouraged by, hyperbolized conceptions of autonomy "conjoined with individualistic conceptions of subjectivity and agency."[23] The self-made achiever is a hyperseparated and hyperbolized autonomous self whose illusion of self-containment is built on denying or backgrounding the contributions of subordinated others and representing the joint product in terms of a hyperbolized individualistic agency whose just deserts are then awarded to "the achiever." In a similar way, global economic systems of property formation are formed in terms of the rationality of the master subject as such an autonomous, separative self. They erase the agency of both social others and of nature, both as land and as preexisting, enabling ecological systems and their products, just as they erase or downgrade the agency involved in so-called women's work.[24]

The increasing gulf in global capitalism between consumption and production and the growing remoteness and irresponsibility of chains of production and distribution institutionalize hegemonic representations of agency in global property formation systems. We can see the same mechanism as that employed by Aristotle at work in current moves to place patented natural organisms under the aegis of intellectual property rights as the creations of reason (assumed to be the identifying property of the center), in which the contributions of other nonhuman systems and agencies are as completely discounted as those of the mother in Aristotle's schema. As far as recognizing the ecological embedment of the dominant culture in the larger system of nature is concerned, current social forms based on hegemonic agency in global capitalism are at about the same stage in their recognition of nature as Aristotelian philosophy was in its recognition of women's role in its account of human reproduction.

When others' agency is treated as background or denied, we give the others less credit than is due to them. We easily come to take for granted what they provide for us and to starve them of the resources they need to survive. This is, of course, the main point of hegemonic construals of agency and labor: they provide the basis for appropriation of the Other's contribution by the One, or center. The "profound forgetting" of nature that ensues from the hegemonic construction of agency, the failure to see otherized nature as a collaborative partner or to understand relations of dependency on it, is the basis of the now-global economic system of self-maximizing economic rationality in which the maximum is extracted and not enough is left to sustain the life of the others on which the rational system is dependent. Hegemonic conceptions of human agency are fostered in the anthropocentric culture of mastery; these are linked to denials of dependency on social and ecological others, which are in turn linked to the application of inappropriate strategies and forms of rationality that aim to maximize the share of the "isolated" self and neglect the need to promote mutual flourishing. Thus the corporate manager is able to assume the contribution of nature in the form of a continuing support base for production, accumulation, and renewal but to deny it in failing to recognize and allow for nature's reproduction and continuation.

Hegemonic constructions of agency that justify appropriation are especially encouraged in the culture/nature dualism typical of Western thinking because its systems of appropriation are based on the idea of applying labor to "pure" nature, as in Locke's argument.[25] The process opens the way for enrichment, but its other side is that the blinkered vision involved is a problem for prudence as well as for justice in the case where the One is in fact dependent on this Other, for the One can gain an illusory and overcomfortable sense of her or his own ontological independence and ecological autonomy. It is just such a sense that seems to pervade the dominant culture's disastrous contemporary misperceptions of its economic and ecological relationships. Countering this denial requires recognition, but "recognition" here must mean much more than just "remember" (as in the case of Mother's Day): recognition means, at least, incorporating the knowledge of their agency into economic institutions and distribution of social resources.

As the forgetting and backgrounding of nature is perhaps the most hazardous and distorting effect of Othering from a human prudential point of view, so the reconception of nature in agentic terms as a coactor and coparticipant in the world and its recognition in distributive terms is perhaps the most important aspect of moving to an alternative ethical framework. Such a reconceived nature would be no mere resource or periphery to our center but another and prior center of power and need, whose satisfaction can and must impose limits on our own conception of ourselves and on our own actions and needs. The nature we would recognize in a nonreductive model is not a mere human absence or conceptually dependent "Other," not a mere precondition for our own star stuff of achievement, but can be seen as an active collaborative presence capable of agency and other mindlike qualities. Nature as biospheric other is not a background part of our field of action or subjectivity, not a mere precondition for human action, not a refractory foil to self. Rather, biospheric others can be other positive presences and ethical subjects to which we can owe a debt of gratitude, generosity, and recognition as prior and enabling presences or ancestors.

Naturalizing and Denaturalizing Strategies

To the extent that, in dominant forms of culture (Western or "modern" culture), nature is treated in terms of this kind of colonizing or Otherizing pattern, its agency, necessity, and contribution, its "labor," tend to be underrecognized, just as women's traditional tasks in household labor and child raising are underrecognized or treated as inessential, as the background services that make "real" work and achievement possible, rather than as achievement or as work themselves. Nature, perceived like woman as an adjunct to the male self rather than as a genuine other, can be represented as inessential and massively denied as the unconsidered background to technological society. Since anthropocentric culture sees nonhuman nature as a basically inessential constituent of the universe, nature's needs can be systematically omitted from its systems of decision making. Dependency on nature is denied, systematically, so that nature's order, resistance, and survival requirements are not perceived as imposing a limit on human goals or enterprises. For example, crucial biospheric and other services provided by nature and the limits they might impose on human projects are not considered in accounting or decision making.[26] We pay attention to them only after disaster has occurred and then only to "fix things up." Where we cannot quite forget how dependent on nature we really are, dependency appears as a source of anxiety and threat or as a further technological problem to be overcome.

How to attribute credit for mixed forms of labor is always a complex matter; think of the problems that can arise in recognizing the contributions of others to an academic paper, for example. But when hegemonic patterns of backgrounding and denial of agency are operative, recognizing contributions and apportioning credit between nature and culture, between the human and the nonhuman, can be especially complex and involve multiple and cross-cutting denials that overemphasize or underemphasize the various elements. The sort of pattern of domination of nature I have outlined has a major bearing on how far and where agency and labor is recognized, as well as on how the structures of denial of agency—human and nonhuman—work and what they are designed to achieve. Generally, in human-centered contexts, the agency of nature is underrecognized, but there is an important class of cases that seem to present exceptions to this rule. I have argued that, in anthropocentric culture, attributions tend to overemphasize the human (especially the privileged human) and underemphasize or deny the agency of nature. But they may also underemphasize or hide the social and overemphasize the natural, for example, in the interests of making outcomes appear less open to change than they really are or from some other motive. This opposing dynamic may derive from the antiliberatory roles of the nature concept, or it may represent the prioritization in particular contexts of some other form of domination—human domination—over the domination of nature.

Numerous examples spring to mind of hegemonic constructions of agency involving nature. Intelligence and other human characteristics that have a substantial relationship to nurture are written down by conservative social forces as hereditary, as nature, in order to give the inequalities in society they are associated with an air of inevitability. As Vandana Shiva points out, corporations involved in genetic engineering patent as nature seed varieties that represent the labor of hundreds of generations of indigenous farmers.[27] Certain kinds of environmentalist foci on ecocatastrophe as a phenomenon of nature preclude any adequate examination of its social aspects and causes.[28]

Another important class of examples concerns the way landscapes are seen as nature, in contrast to culture. Thus Kate Soper (1994) points to the failure to recognize the labor of otherized human groups (the laboring people) and the human social relations that have gone into places now presented as nature, for example, the countryside of England. In Australia, the colonizers denied the possibility that the indigenous inhabitants, who were seen as semianimal, could have ecological agency, and landscapes that often had substantial indigenous inputs and management were taken to be in "the pure state of nature," including no element of indigenous human labor in their formation. The Blue Ridge Parkway in North Carolina now winds its way through surrounding land that is increasingly suburbanized, but planning codes require that buildings be carefully hidden from the sight of vehicles on the parkway so that the illusion that it travels through wilderness can be maintained for the benefit of the tourist industry. To consider another example, cosmetic strips of unlogged forest along highways in logging areas are often used to hide destructive logging activities and, as in the case of the Blue Ridge Parkway, give the impression that there is much more nature around than there actually is so that destruction of the remainder can continue without objection or hindrance.

In the case of deceptive naturalness, describing something as nature tends to be not so much a way to overacknowledge the contributions and workings of nature as a way to underacknowledge the human activities and social relations involved and the extent of prior ownership or human construction. In these contexts, we may need to denaturalize, to demote or supplement, the emphasis on nature and note the presence of human influences, which have been hidden, although this will rarely involve a complete denial of the influence of nature. Although these cases seem to be an exception to our general claim that, in dominant anthropocentric culture, nature's influence has been denied in favor of overcrediting the human, in fact, they involve a more complex, multiple set of denials registering multiple forms of oppression and colonization. We need a complex, case-sensitive response to these complex denials, involving both naturalizing and denaturalizing strategies in combination.

We can sum up some of the complex classes of cases and the corresponding strategies required to counter these denials as follows:

Type 1: Naturalizing 1 (deceptive naturalness 1). Counting something as nature in the sense of "pure nature," when it in fact has a human contribution (not merely a human influence), hides or denies the human social relations that have gone into that construction, often in the interests of making it seem unchangeable, of appropriating it, or for some other deceptive purpose such as suggesting there is more of it than there is. Examples: gender oppression ("woman's nature"), the Blue Ridge Parkway, *terra nullius*. In contexts of oppression, an unchangeable "woman's nature" is often attributed to women. Deceptive packaging of a road (such as the Blue Ridge Parkway) hides human habitation in order to suggest there is more nature than there is. In *terra nullius*, the extent of indigenous influence on a landscape is played down in order to justify seizure and appropriation of land as "nature." We need strategies of denaturalizing, that is, recognizing human agency as well as that of the nonhuman sphere, in response.

Type 2: Overhumanizing (deceptive humanness). Counting something (e.g., a place) as purely human when it involves the labor of nature jointly with human labor can hide or deny the ecological dependency relations in that construction. This is the dominant position, because, as we have seen, nature's operations and contributions to our joint human-nature undertakings are overwhelmingly denied or backgrounded in Western culture. We need strategies of naturalizing in the sense of recognizing nature's agency, for example, as in acknowledging and providing for the continuation of "ecosystem services."[29]

Type 3: Naturalizing 2 (deceptive naturalness 2). Given the structure of type 2, one common way to hide certain human social relations and contributions (e.g., to a place) is to count the human groups involved as nature, so that their contributions will not need to be credited or noticed. In this case, too, we need to respond by denaturalizing, in the sense of distinguishing the human groups concerned from nature and showing how their role has not been credited. But at the same time we need to naturalize, to credit the nonhuman agency that has not been credited.

Some groups historically identified with the body and the animal, such as indigenous people, women, and those who do manual labor, are especially likely to have the outcome of their labor represented as nature rather than as mutual construction between humans and nature. Such a hegemonic construction of agency based on associating these groups with nature seems to be what lies behind the case of patenting seeds, the case of indigenous people in Australia, and the case of the agricultural workers whose bodily labors over generations helped form the countryside now seen as nature. The basic motivations for such denials of their contribution is clear: it opens the way ethically for appropriation by the more powerful or prestigious of what the Others have helped create. Thus Australia was seen as *terra nullius*, the land of no one, open to appropriation because indigenous people were counted as semianimal nomads and their ecological agency in and attachment to the land discounted.

It is important to note that this strategy relies on discounting the agency of the nonhuman sphere, that is, nature itself. It has been possible to discount the agency of subordinated groups of humans by counting them or their agency as nature only because nature's agency is itself normally denied and backgrounded in Western culture. Soper problematizes cases of type 1 but none of the remainder and gives us an inadequate sense of our embeddedness in nature by failing to problematize cases of types 2 and 3. Cases of type 3 make up an important class of cases where the agency of certain groups of humans in the land is hidden, but we cannot understand type 2 cases without understanding type 3 cases. We can be grateful to Soper for clarifying cases of type 1. But we still need to take account of the other two types, and this means understanding and countering the dominant tradition's denial of recognition to nature and nature's agency. Our countering of cases of deceptive naturalness needs to be balanced by countering cases of deceptive humanness.

Dispensing with Nature?

One of the problems with such a partial account as Soper's, which focuses only on type 1 cases of deceptive naturalness unbalanced by type 2 cases of deceptive humanness, is that a common response to demoting or supplementing nature's agency in cases of deceptive naturalness is to suggest that maybe nature is something illusory we can dispense with or dismiss, that nature is not really other at all but is entirely constructed by us. Generalizing from particular cases of deceptive naturalness to cast generalized doubt on nature and nature's agency is a major basis for nature skepticism and constructivism applied to nature. But now, of all times, when we press so many natural limits and most need to be aware of what we are destroying in the nonhuman sphere and how dangerous this is for us, skepticism and constructivism of this generalized variety is immensely problematic, since we cannot come to terms with another whom we do not recognize as presenting to us any independent form of agency or limit on our actions. Once it is realized that cases of type 1 are only part of the story, the sorts of skepticism and constructivism that would dismiss concepts of nature and nature's agency in a completely general way are no longer so attractive. A more constructive response to these complex recognitions and emphases might be to develop a larger politics of place that could recognize both human and nonhuman agencies—including the "labor" of earth processes—together with the constraints imposed on each by ecological relationships, as all inscribed in place.

There are however other, rather better, reasons for such generalized skepticism about nature. Many objections to the concept of nature assume conceptions of nature and culture drawn from anthropocentric dualisms that treat humanity and its cultural products as emphatically distanced from a nature defined in hyperseparated terms as pure, undefiled by human interference. Such polarities obscure or problematize the interactions and meeting points between nature and culture. In the hyperseparated picture, it appears that only "pure nature" is nature and that nature must be a realm totally separate from the human. The nature skeptic then objects that is impossible to find (especially nowadays) forms of the nonhuman that do not carry some human influence and even that there is no such thing as natural air in the sense of air that has never been breathed by anyone else. But if air is not nature (or natural), what is? The nature skeptic then concludes that there is no such thing as nature, since nothing can be guaranteed to be without human influence.

This kind of argument has many flaws. One does not have to abandon the claim to be another, separate person just because someone else influences or has some impact on one. So why should we have to abandon the claim that there is another that is nature just because it often carries some human influence? To be other is not necessarily to be purely other. Something does not cease to be or to involve nature just because it has elements of the human, shows some human influence, or is interpreted by humans through human culture.[30] We have to remember that these dualistic, polarizing senses of nature that are heavily implicated in the current wave of nature skepticism are not the only possible or only current senses. Nevertheless, since the term "nature" is readily subject to this type of ambiguity, we should, where there is a risk of miscommunication, look toward a range of other expressions that are not so easily open to the assumption of purity. For example, we might speak of "the earth" or of ecological or biospheric systems or processes, although this can require from us a degree of specificity we may not always or easily be able to supply.

In arriving at a just assessment for crediting mixed forms of agency, we have to decide what sorts of influences to emphasize over what others and by how much, and this can be a judicious and difficult matter, always presenting some element of the political in reflecting whose stories are told, whose efforts remembered. We have to consider, too, a range of cases, not just those we have experienced ourselves and can see near where we happen to live. The earth is a big and still very diverse place. Many ideas of nature, especially in a place like Britain, seem primarily to aim to draw a contrast with urban or domestic existence, nature as the countryside, an area whose long history of human influence gives talk of constructing nature greater credibility than it would be likely to have on some remote southern glacier. We also have to consider those cases where human influences are very slight, and those accounts that do not consider this may be unsuitable for many non-European contexts. Philosophical concepts and terminology need to be sensitive to our present problems and context, and it thus seems misleading at best to talk about humans constructing nature in any general way. To talk of construction is in many contexts to imply that what is often mere influence or impact is actually control, to suggest that because we humans have an (often blind) impact or effect on the biosphere, we can produce the outcome we want. It is also to suggest that we can *re*construct it, when we cannot even reconstruct a bird's feather. Both these terms can involve serious overestimations of human contributions in a range of cases and invite slippery slides into implications of control that are very dangerous in the present circumstances.

An alternative image sometimes proposed to supplant that of construction as a metaphor for human interaction with and influence on nature is the closely related metaphor of production, warranted according to some because "nature bears the indelible trace of [human] labor."[31] Human construction of nature was seen to imply skepticism because, in the hyperseparated concept, to be nature was to be completely independent of the human, so what is not fully other cannot be nature. On the surface, productivism does not appear to imply skepticism, since rather than insisting that nature does not exist, productivists talk happily about our producing nature. Nevertheless, productivism will produce the same skeptical outcome as constructivism unless the hyperseparated concept of nature is abandoned, since where nature is defined as completely other to human, a nature produced by humans cannot really be nature. But once we abandon hyperseparation, we do not need anything as overgeneralized, one-sided, and monolithic in its recognition of agency as productivism; we can often make do with more pluralistic and context-sensitive concepts of influence, interaction, and mutuality instead. These concepts connote some degree of independence, which has been felt to be problematic mainly because independence and otherness is mistakenly equated with hyperseparation. They are compatible with recognizing nature as a sphere of agency and coagency that is distinct but not hyperseparate.

The concept of production has many drawbacks as a contender for the job of general model for human relations to nature and to the nonhuman sphere (if such a general model is indeed possible; perhaps all models should acknowledge their incompleteness and partiality): it continues the modern anthropocentric tradition of denying nature's agency, placing the human on the active side and the nonhuman on the passive, instead of allowing for the possibility of equal and mutual distribution of activity and passivity. It recognizes labor and agency in a hegemonic way as occurring on just one side, the human one, and blocks any conceptualization of nature's labor in the production of what is sometimes described as ecosystem services. Static, one-sided narratives of humans producing (managing?) nature do not envisage the possibility of mutual production, failing to allow for humans coproducing with an equally productive nature or to recognize that nature also produces us as well as we it. They continue to suggest human control to at least the same degree as the concept of construction and seem especially inappropriate for those many cases where the outcome is neither planned nor anticipated by those responsible for it.

The productivist's hyperbolized concept that humans produce alternative natures could be restated in more modest and less misleading terms as the idea that our actions can contribute (often unwittingly) to bringing about alternative forms or states of nature. Concepts of influence on and interaction with nature connote some degree of independence, which the productivist as well as the constructivist feel to be problematic, mainly, I think, because the difference between separation and hyperseparation, and therefore between independence and pure otherness, has not been clearly registered. According to Neil Smith, a further virtue of the production model is that it is a positive concept, but a closer focus reveals that this positivity is at least in some cases illusory since the positive activity of humans involved in producing, say, a desert may in fact be that of destroying life and complexity.[32] As a general model for human relations to nature, the production metaphor vastly overstates human causal contributions in an important range of cases. The productivist's answer to the question, how exactly *did* we produce that mountain range? is unclear but seems likely to reduce to little more than some version of the influence thesis.

Finally, productivist emphasis on labor as the central feature of the human is not culture-neutral but is especially associated with industrial modernity, Western culture, and its insistence on human labor rather than the agency of nature as the source of wealth and life. If Western culture has made human labor and humanizing transformation of the land the source of human ownership, there are other cultures for which place-based identity and narrative, together with the naming of the land, create title, in the sense of the right to live on the land and gain livelihood from it.[33] The metaphor of production, with its insistence on the centrality of human labor, intensifies rather than reduces the problem of Eurocentrism and androcentrism in the concept of nature,[34] as well as the problem of anthropocentrism or subject/object thinking in leftist theory.[35] With its implications of control, it amplifies the problems of anthropocentric backgrounding and its consequences in neglect, indifference, blind spots, and overconfidence.

Donna Haraway suggests that nature is now old hat, that we have moved past the time when the concept is useful.[36] This postnaturism seems to me to deserve the same retort as the similar advocacy of postfeminism: the fact that a few people have begun to contest the devaluing and agentic disappearance of nature or woman does not mean that we have arrived at a system of thought or life that can dispense with the concept. Just as we have a long way to go to reach postpatriarchy, we have a long way to go in recognizing and consciously maintaining the ecological relationships on which human culture depends. The concept and experience of nature are needed to make these relationships more apparent to people living increasingly urbanized lives in what they think of as culture, a sphere often but mistakenly seen as of exclusively human construction and agency.[37] Countering hyperseparation and false polarization between nature and culture through recognizing hybridity and continuity are important projects, but they do not imply reducing both binary terms (nature and culture) to a single term (culture)—an unsatisfactory formula for resolving dualistic constructions, especially where the victorious survivor is the traditionally dominant one in the problem narrative. A more complete shake-up of the dualistic boundaries would involve working out new terms for *both* the relata, both nature and culture.

As part of escaping dualist construction, we certainly should reject the idea that culture is self-enclosed and nature purely other (which seems often to be part of what people who talk about dissolving boundaries have in mind). But rejection of these polarizing constructions of nature does not require dispensing with nature, prioritizing culture over nature, or absorbing the one pole of nature into the other pole of culture. Such reductionist measures are not helpful reinventions of the concept of nature; rather, they move with the mainstream of the Western tradition of anthropocentrism and backgrounding. They support rather than disrupt the modern sense, especially fostered in urban life, that we humans are completely immersed in a self-enclosed sphere of our own that we can call "culture."[38] Indeed, the central problem can be taken to lie just as much in this concept of culture as in the concept of nature itself. The idea that human life takes place in a self-enclosed, completely humanized space that is somehow independent of an inessential sphere of nature, which exists in a remote space "somewhere else," might be seen as the foundational delusion of the West. A dangerous doctrine, strongly implicated in the environmental crisis, this framework of self-enclosure is the love child of the old dominant narrative of human mastery and centrality mated with the much younger circumstance of human experience of commodification in the global city. We augment rather than disrupt this foundational delusion by adopting what amounts to an intensified form of nature denial.

Inclusive Accounts of Nature

None of this is to say that the concept of nature (or indeed of culture) is in order as it stands. To the extent that nature in the West has been defined in and through anthropocentric, androcentric, and Eurocentric narratives, justifying the dominance of humanity and mastering reason over the nonhuman sphere, it is an important project to problematize the concept of nature as so defined. But there is no necessity for concepts of nature to be trapped inside these stories. Concepts of nature are like those of woman: they can be liberated through imagining and implementing new narratives. To the extent that they have been forged in a dualistic structure of contrast to the human and to reason, we must constantly problematize their old meanings and criticize their old stories. But neither for women nor for nature is it helpful to demand abandonment, as opposed to reworking, reinventing, reimagining, and to neglect the option of making new stories.

Against those who see the role of nature in these guiding narratives as a reason for totally rejecting any nature concept, I would argue that abandoning concepts of nature at this point for this reason could leave the dominant narratives unresolved but still influential in their background status, while the failures of knowledge and rationality they inspire become increasingly critical. Further down the track, I hope, we will no longer place such weight on the concept of nature as a catch-all concept defined in contrast to the human, favoring more multiple and less homogenizing contrasts and alliances. But a society inured to mastery and pressing its ecological limits desperately needs ways to acknowledge elements and limits of otherness in the sphere of the nonhuman, since final conquest of that otherness is incompatible with ecological survival. Such a society therefore requires some concept of nature, whatever the language used to express it.

We can't just strike out with a brand-new story with brand-new characters, or no story at all, and hope to make sense of where we are. For better or worse, the old story holds the keys to who we are, why we are here, where we have come from, and where we might now go. Reflecting, in a culturally self-critical vein, on the master narratives of humanity, culture, reason, and nature in the West can give us valuable clues as to why the dominant forms of "developed" society, and the relationships with nature they have built on a form of denial, are now failing the most basic tests of rationality and fitness for survival.[39] Reflecting on that failure can suggest some guidelines for devising counterstories that might disrupt the ideals and projects of mastery of the old (although I am not suggesting that all our narratives and traditions need to be abandoned).

For example, it seems that much more is required to disrupt the old narratives than celebrating and fostering the breakdown or blurring of boundaries between nature and culture, the main strategy for rethinking suggested in the work of many postmodern thinkers such as Donna Haraway.[40] A generalized strategy of boundary breakdown is a shallow and imprecise plan for resolving dualistic constructions, since boundary breakdown is an ambiguous feature that can occur in oppressive as well as liberatory ways. Although reclaiming the denied elements of continuity and overlap between nature and culture is crucial to resolving dualized construction and hyperseparation, certain kinds of boundary breakdown imply lack of respect and are implicated in projects of colonizing and erasing the other. A colonizing consciousness aims to form hegemonic wholes that involve the dissolution of the boundaries and integrity of the colonized other, resulting in projects of assimilation and cultural destruction. Placing a destructive mining venture in a great wilderness area may help break down the boundary between nature and culture but is not a cause for celebration or a useful disruption of a damaging tradition. If some kinds of boundary breakdowns are *with* power and only certain kinds are *against* it, understanding the difference requires a theory of oppression that will take us well beyond the fashionable preoccupation with breaking boundaries.

There are several important theoretical challenges, so far not much explored, in countering the West's foundational blind spot, the delusion of culture as a self-enclosed space hyperseparate from an inessential nature. One of these is to give an inclusive rather than an exclusive account of the human/nature relation, for if nature is the sphere of the nonhuman, the further fact of the inclusion of humanity in nature means that this "not" must be read in an inclusive rather than exclusive way. That is, nature as ecological process should be seen as a larger sphere that takes in but greatly exceeds the human: it is the "more-than-human,"[41] in contrast to the centric and oppositional account of it as alien or lesser. One source of the view of nature as alien and lesser is the dualistic partner concept of the human I have discussed elsewhere under the heading of the Differential Imperative.[42] This type of account emphasizes as authentically human, and as the human ideal, those features that supposedly make humans different from the animal and from the larger natural world, rather than what is shared with them. Such an exclusionary understanding of the human contributes in a major way to the sense of humans and human culture as outside nature. The challenge to dualistic accounts of nature involves rethinking both sides of the oppositional human/nature contrast.

An associated task is to give a thorough account, not only at the ecological but also at the conceptual level, of how culture might be included or embedded in nature (as ecological process) and of why this very basic fact about the world is now seemingly invisible to so many of us. In at least one important sense of nature, the relationship of nature to culture involves a determinable/determinate type of logical relationship. For example, we seem to locate hunger and food in nature, but spaghetti sauce and the desire for it in culture, where spaghetti sauce is a determinate of the determinable concept of food.

The determinate, food, is of the realm of necessity, but the determinable is contingent, the demand for it a matter of individual and cultural choice. We and all other animals need to eat food as a matter of (our) biological nature, of necessity, but the choice of what we eat is part of the realm in which we (and all other animals) exercise freedom and create culture.

If this is so, the embedment of culture in nature turns in part on logical differences in kinds of narratives linked through different levels of generality and specificity. We can vindicate here a concept of nature as a participant in all we do in culture, so that the foundational delusion of human self-enclosure can be seen to involve not just an ecological but also a logical level of failure. Perhaps more important for the crucial questions of social change, the neglect of the embedment of lived experience in this higher-order level of physical generality in contemporary life also involves failures at the level of social organization. Filling out at this level an account of embedment in nature and its invisibility and neglect in contemporary culture would require reference to contingent factors, such as the increasing irresponsibility and remoteness of chains of production and consumption in global markets and the forms of culture and urban/rural experience they select and support.[43]

But this foundational delusion of cultural self-enclosure is also the delusion of the autonomous center shaping hegemonic concepts of agency and achievement, and concepts of culture and of nature are complicit in it to the extent that they fall in with the dominant anthropocentric meanings the centric narrative gives them. One of these centric meanings is to think of nature always in the negative, as an absence of the qualities of the human, of the center. Here, the dominant meaning of nature as the nonhuman sphere would seem to be deeply problematic. To counter the foundational delusion, we need to think of nature as a positive presence, or as a community (positive presence) of positive presences, and not as a failure to be or to involve the human. Does this mean that the very use of the negative term "nature" as meaning "the nonhuman" is inevitably anthropocentric? I think it is a little more complex than that.

There is a serious problem about how to fix in any absolute way the idea of a negative term in view of widely accepted transformation principles such as double negation. As I argue in Plumwood 2001, the use of a term explicitly employing or defined by negation, whether "nonhuman" or "other-than-the-human," is neither a necessary nor a sufficient condition for centric and oppressive uses or for concluding that "nonhuman" cannot designate a positive presence. It rather depends on whether that term is used to conform to a centric pattern that brings all meanings back to a privileged and exclusionary center—of meaning, experience, or value. "Nonhuman" may in some contexts be used in this way, but in other contexts it may not. There is no immediate, general, or context-free way to decide whether a term like "nonhuman" is used in centric ways or not, short of determining its pattern of use. But if "nonhuman" is not automatically anthropocentric, we can at least say in what pattern of attributions it would be: when matters are constantly brought back to questions of otherness, sameness, or difference from the human as center. To define nature as a lack of human qualities, for example, is not only to deny continuity and overlap but to define it both as inferior and always in relation to the human as center.

As part of breaking up the centrically polarized configuration of the human/nature contrast, we need some demassifying of the concept of nature, which, as part of this opposition, tries to cover the whole nonhuman sphere in a homogeneous way. I see as a clear symptom of human centeredness the obsessive focus on the kind of configuration of nature that gives a homogenized contrast with the human such a central cultural role. Where our frameworks of thought are no longer human centered, we should be able to break up this dualized configuration of the human/nature contrast to establish a more fully interspecies ethic that draws contrasts in many different and diverse places and ways. When we are in a position to decline the discourse of human centeredness and instrumentalism, we will be in a position to cease the excessive focus on this major simplifying and homogenizing contrast of human versus nature and to reconfigure the field in terms of a more complex and varied set of interspecies distinctions and alliances to create an also more complex and less centric configuration of the world.

But, contra Haraway, I would say that while we still have mainly to counter and encounter the dominant story in which the world is always referred back to a human center and is configured mainly as a human resource, we are obliged to continue some of the major counter focus on the other side of this contrast, on nature, at least in an important range of contexts. This is so because, where human centeredness and oppositional practice is still at work, there is a need for some forms of counteraffirmation of the devalued elements configured as nature (although these should not take the form of reverse centrism or polar reversal) that make the objective of immediately and finally breaking up all contrasting configuration unrealistic and unwise.

One ultimate aim of reworking nature concepts should be an interspecies politics and ethics that ventures beyond the polarized configurations that classify the world into contrasting sides of human and other or alternatively in terms of human and similar (its hegemonic variant in the currently popular moral extensionism of contemporary philosophy, which so neatly confirms philosophers' intuitions of human superiority). Brian Luke has provided some interesting examples of ethical configurations that avoid the old human-centered species hierarchies and oppositions built on the species generalities of the "great-chain-of being."[44] This set of human-centered ethical configurations assumes a descending order of species, usually based on degree of similarity to the human in some selected area, and attempt to resolve interspecies conflict by focusing on a few general qualificatory properties of species that are supposed to pick out ethical winners and losers.[45] This sort of moral extensionism is hegemonic, bringing all valuations back to similarity to, or difference from, the human as the norm or center, and it is also vastly oversimplifying. Without its pervasive modeling of ethical relations in species terms of human and other, we can aim at a form of ethical consideration that is more contextual and can recognize unassimilated otherness. Once these dualized configurations have been broken down, as they can be in many contexts, questions of species differences (in the sense of being other to the human) and hierarchies of similarity to the human (e.g., consciousness) need not loom so large in the formulation and application of ethical principles.

The interspecies politics that the breakdown of dualized configurations can make possible could make alliances across species that deemphasize the importance of generalized and stereotypical species frameworks and differentials and open the door to new kinds of communicative experience, new, that is, for Western culture, which might just begin to frame the world in more sensitive and nuanced terms than we can imagine while wearing the simplifying blinkers of human superiority. These reframings prepare the ground for movement from monological and dualistic types of relationship with nature toward the kinds of structures of relationship we need to develop in order to begin addressing the environmental crisis at the level of culture. They can open the way for a culture of nature that allows for much more in the way of contextual and negotiated relationships of communication, balanced dialogue, and mutual adjustment among species, starting with our own, in what would be, in the old terms, a liberatory blending or meeting of nature and culture.

NOTES

An earlier version of this essay appeared in *Capitalism, Nature, Socialism* 12, no. 4 (2001): 3–32.

1. On this issue and on the dominant narrative, see Plumwood 1993, Merchant 1980, 1996.
2. See Ruether 1975. As Soper 1994 remarks, "all Western discourses on nature, including those most critical of its abuse, carry with them the ethnocentric legacy of a metaphysical tradition that has covertly identified the 'human' side of the humanity-nature distinction with 'civilized'/'developed' humanity" (p. 10). The same can be said, of course, of nature's dualization opposite reason, which, how-

ever, rarely receives the blame due to it in this connection. Both "nature" and "human" concepts have involved the formation of hegemonic wholes.

3. See Spretnak 1997.
4. Prokhovnik 2000.
5. Progress can even be linked to the reworking of that narrative. For example, one aspect of progress might be seen as success in decolonizing and decentering concepts of the human and of reason.
6. See Plumwood 1981.
7. For a discussion of this type of case, see Plumwood 1993 and 1999.
8. See Benton 1990.
9. See Plumwood 1999.
10. This argument is put forward by Dennett 1996 as a reason for not letting too many animals into the rights club. For critique, see Plumwood 1998b and 1999.
11. See Ruether 1975, Spretnak 1997.
12. For a more detailed argument that John Locke's formula subsumes the agency of nature, see Plumwood 1999.
13. It follows that "white-supremacist capitalist patriarchy" remains an incomplete specification, but the idea of identifying such a system by giving a list is in any case severely problematic. See Plumwood 1996.
14. See Benton 1993.
15. Smith, 1996 p. 46 n.
16. See Plumwood 2000.
17. For a summary, see Plumwood 1999.
18. On such a concept of "race," see Gould 1981.
19. For examples, see Said 1979.
20. In some hybrid forms, the key assumptions of discontinuity may take a more subtle form; thus, in Roman Catholicism, continuity is admitted at the level of the body but denied at the level of the mind or spirit, considered to be the level that defines the truly human.
21. See Plumwood 1993.
22. See Marglin and Marglin 1990.
23. Code 2000, p. 184.
24. See Waring 1988.
25. For details. see Plumwood 1998b.
26. See Waring 1988.
27. See Mies and Shiva 1993.
28. For a critique of this hyperseparation, see DiChiro 1998.
29. Although the concept of "ecosystem services" can be dangerously human-centered if it fails to recognize that such services have a much wider range of beneficiaries than the human and if it supports instrumentalizing and servantlike conceptions of the nonhuman sphere.
30. See Plumwood 1998b.
31. See Smith 1996, p. 52.
32. Here, as in many other places, the concept of negative and positive concepts and properties is highly problematic. See Plumwood 2000. Michel Foucault's use of

production as the leading metaphor for the action of power suffers from many of the general difficulties of production, as well as depending, like the concept explicated by Smith, on the problematic positive/negative property distinction I discuss in Plumwood 2003.

33. See Snyder 1990, p. 7.
34. Of course, productivism in itself has been heavily criticized as associated with industrial models of society, with workerism, and with androcentrism. See Benhabib and Cornell 1987.
35. For a recent critique of this aspect of Marx's thought, see Brennan 2000.
36. Haraway 1997.
37. Nature/wilderness can be captured by or made complicit with this delusion—especially when it is treated as a sphere apart—but it also has the potential to unsettle and counter it through the revelations of the radical dependency of human life that wilderness journeying can bring.
38. The postmodern tendency to reduce nature to culture is encouraged by the assumption that nonhuman nature can be treated in just the same way as the human body (see, for example, Prokhovnik 2000), despite the fact that the human body is clearly much more closely integrated in human culture than is the nonhuman sphere.
39. An important group of these narratives is discussed in Merchant 1996.
40. The overemphasis on this strategy seems to arise from the familiar conflation of separation and hyperseparation. Although Haraway in some places clearly recognizes nature as an active, independent agency (e.g., Haraway 1988), this sensibility is less in evidence in later work such as Haraway 1997. Her concept of nature as "a body-in-the-making" in Haraway 1997 seems ambiguous between the idea of nature as a "vacant lot" or as a "development proposition/potential"—a neutral field passively open to the inscriptions of culture—and the determinable concept I outline below.
41. In David Abram's terms. See Abram 1996.
42. See Plumwood 1993.
43. The concept of remoteness is discussed in Plumwood 1998a.
44. Luke 1995.
45. These configurations neglect context and display what Luke calls "generalism."

REFERENCES

Abram, David. 1996. *The Spell of the Sensuous: Perception and Language in a More-Than-Human World.* Pantheon, New York.

Benhabib, Seyla and Drusilla Cornell, eds. 1987. *Feminism as Critique.* Polity, Cambridge, UK.

Benton, Ted. 1990. "Humanism = Speciesism? Marx on Humans and Animals." In Sean Sayers and Peter Osborne, eds., *Socialism, Feminism and Philosophy: A Radical Philosophy Reader*, 235–274. Routledge, London.

——. 1993. *Natural Relations: Ecology, Animal Rights and Social Justice.* Verso, London.

Brennan, Teresa. 2000. *Exhausting Modernity: Grounds for a New Economy.* Routledge, London.

Code, Lorraine. 2000. "The Perversion of Autonomy and the Subjection of Women." In Catriona Mackenzie and Natalie Stoljar, eds., *Relational Autonomy: Feminist Perspectives on Autonomy, Agency, and the Social Self,* 181–212. Oxford University Press, Oxford.

Dennett, Daniel. 1996. *Kinds of Minds.* Weidenfeld and Nicholson, London.

DiChiro, Giovanna. 1998. "Nature as Community: The Convergence of Environment and Social Justice." In Michael Goldman, ed., *Privatizing Nature: Political Struggles for the Global Commons,* 120–143. Pluto, London.

Gould, Stephen Jay. 1981. *The Mismeasure of Man.* Norton, New York.

Haraway, Donna. 1988. "Situated Knowledges: The Science Question in Feminism as a Site of Discourse on the Privilege of Partial Perspective." *Feminist Studies* 14 (3): 575–599.

——. 1997. Modest_Witness@Second_Millennium. Routledge, London.

Haraway, Donna and David Harvey. 1995. "Nature, Politics, and Possibilities: A Debate and Discussion with David Harvey and Donna Haraway." *Society and Space* 13(5): 507–527.

Luke, Brian. 1995. "Solidarity Across Diversity: A Pluralistic Rapprochement of Environmentalism and Animal Liberation." *Social Theory and Practice* 21 (2): 333–358. Also in Roger S. Gottlieb, ed. 1997. *The Ecological Community.* Routledge, London.

Marglin, Frederique Apffel and Stephen Marglin. 1990. *Dominating Knowledges: Development, Culture and Resistance.* Oxford University Press, Clarendon, Oxford.

Merchant, Carolyn. 1980. *The Death of Nature.* Wildwood House, London.

——. 1996. *Earthcare: Women and the Environment.* Routledge, London.

Mies, Maria and Vandana Shiva. 1993. *Ecofeminism.* Zed Books, London.

Plumwood, Val. 1981. "On Karl Marx as an Environmental Hero." *Environmental Ethics* (fall): 237–244.

——. 1993. *Feminism and the Mastery of Nature.* Routledge, London.

——. 1996. "Anthropocentrism and Androcentrism: Parallels and Politics." *Ethics and the Environment* 1 (fall): 119–152.

——. 1998a. "Ecojustice, Inequality and Ecological Rationality." In John Dryzek and David Schlosberg, eds., *Debating the Earth: The Environmental Politics Reader,* 559–583. Oxford University Press, Oxford.

——. 1998b. "Intentional Recognition and Reductive Rationality: A Response to John Andrews." *Environmental Values* 7 (December): 397–421.

——. 1998c. "Wilderness Skepticism and Wilderness Dualism." In J. B. Callicott and M. Nelson, eds., *The Great New Wilderness Debate,* 652–690. University of Georgia Press, Athens.

——. 1999. "Ecological Ethics from Rights to Recognition: Multiple Spheres of Justice for Humans, Animals, and Nature." In Nicholas Low, ed., *Global Ethics and Environment,* 188–212. Routledge, London.

——. 2000. "Integrating Ethical Frameworks for Animals, Humans and Nature: A Critical Feminist Eco-Socialist Analysis." *Ethics and the Environment* 5 (3): 1–38.

——. 2003. "Feminism and the Logic of Alterity." In Marjorie Hass and Rachel Falmagne, eds., *Feminist Approaches to Logic.* Rowman and Littlefield, Lanham, Md.

Prokhovnik, Raia. 2000. *Rational Woman: A Feminist Critique of Dichotomy.* Routledge, London.

Rogers, Raymond. 1995. *The Oceans Are Emptying: Fish Wars and Sustainability.* Black Rose Montreal.

Ruether, Rosemary Radford. 1975. *New Woman, New Earth.* Seabury, Minneapolis.

Said, Edward. 1979. *Orientalism.* Vintage, New York.

Smith, Neil. 1996. "The Production of Nature." In George Robertson, Melinda Mash, Lisa Tickner, John Bird, Barry Curtis, and Tim Putnam, eds., *FutureNatural: Nature, Science and Culture,* 35–54. Routledge, London.

Snyder, Gary. 1990. *The Practice of the Wild.* North Point, New York.

Soper, Kate. 1994. *What Is Nature? Culture, Politics and the Non-Human.* Blackwell, Oxford.

Spretnak, Charlene. 1997. *The Resurgence of the Real.* Addison-Wesley, Reading, Mass.

Vogel, Steven. 1996. *Against Nature.* State University of New York Press, Albany.

——. 1997. "Habermas and the Ethics of Nature." In Roger S. Gottlieb, ed., *The Ecological Community,* 175–192. Routledge, London.

Waring, Marilyn. 1988. *Counting for Nothing.* Allen and Unwin, Auckland.

THREE

Is Nature Autonomous?

KEEKOK LEE

THE ANSWER to the question posed by the title of this essay is "yes." I begin by distinguishing various senses of "nature" and clarifying the sense in which nature may be said to be autonomous. After this, I argue that autonomy should be defined, on the one hand, negatively as what exists and continues to exist independently of human intentionality, control, manipulation, or intervention and, on the other hand, positively in terms of being self-generating and self-sustaining and explain why the term may be applied to biotic as well as abiotic nature. I conclude by showing that autonomous nature is more than a social construct.

Defining Nature

The term "nature" is by no means self-evidently clear. Its unclarity concerns both its sense and its reference. First, let us consider its reference. Some who bemoan the end of nature, like Bill McKibben, seem to have assumed that it refers to (our) Earth only or, rather, to its surface. But this assumption, made in McKibben's *The End of Nature,* is too simplistic.[1] There is more to Earth than what he assumes, as we shall see.

The term also refers to the entire universe out there or, at the very least, for the purpose of this discussion, to the solar system of which our planet is a member. This is but a statement of plain fact and is therefore not expected to be contentious. However, apart from the actual celestial bodies and the objects on them, such as mountains, frozen lakes of carbon dioxide, or organisms, one must not forget that the term indicates as well the various natural processes—physical, chemical, hydrological, biological—that are continuously taking place.

In arriving at the denotation of the term, which McKibben has wrongly assigned only to the surface of Earth, he also seems to have been affected by a particular connotation he has assumed, namely, that for something to count as part of nature, it must have escaped human impact. On Earth's surface, apart from Antarctica, it is difficult to find any region that does not bear substantial, if not overwhelming, human impact, such as permanent habitation, deforestation, cultivation, mining, and so on. This difficulty is compounded by the fact that industrialization has ensured that its pollutants are carried to many relatively uninhabited parts, including even Antarctica. Using such implicit criteria about the denotation and connotation of the term, it is not surprising, then, that McKibben provocatively proclaims the end of nature.

The foregoing clarification casts doubt on his assertion, however. For instance, the depths of Earth's oceans are relatively untouched, and apart from some drilling, to a few kilometers deep into its mantle, its core has not even been explored directly via instrumentation. Of the other bodies in our solar system, a few humans have walked on the moon, some machines have landed on Mars, for example, and some space vehicles, with or without people, have orbited or are still orbiting the other planets. Some machines that landed were left behind. Bits of others are still orbiting in space, and still others have been burnt up, thereby leaving some human imprint in deep space. This imprint may be said to be relatively ephemeral, unlike the impact made on Earth's surface, where humans clear forests or flatten hills, if not mountains: such activities involve transforming what is natural into what is artifactual.[2]

It may be wise to distinguish at least seven senses of the term "nature":

1. Nature in the cosmological sense. Whatever can be identified in terms of spatial and temporal coordinates falls within it; its antonym is the supernatural. In this sense, the Great Wall of China is as natural as the Great Barrier Reef; goddesses and angels or heaven and hell, however, are not. This sense embraces all humans and all nonhuman entities that exist or have existed in the universe.
2. Nature as pristine nature: that which bears no human imprint or impact.
3. Nature that bears or has borne human impact, the antonym of nature as pristine nature.
4. Nature as natural kinds, such as gold.[3]
5. Nature as elements in the periodic table, such as hydrogen.
6. Nature as nonhuman nature, including both the abiotic and the biotic, as well as the processes that produce and sustain the abiotic and biotic elements.[4]
7. Nature as the contrast or foil to human artifacts.[5]

To assess which of the senses identified above would be relevant to the issue under discussion, one needs to draw attention to the question itself—is nature autonomous?—and to show that behind it stand really two questions: "autonomous, but of what?" and "autonomous, but in what way?" I give a quick answer to the first here (the second will be examined in detail later). There are two possible answers: autonomous of God/other supernatural entities and autonomous of humans. One takes it that the task here is not a theological one. But if nature may be said to be autonomous of humans, then the first sense (nature in the cosmological sense) may be ignored, as it would be too wide or inclusive. Adhering to it would preempt the kind of discussion of interest here, which explores the possibility that nature, in some senses other than the cosmological, could be said to be autonomous. Furthermore, nature could be said to be autonomous in the sense that it follows its own trajectory (or, more accurately, different parts of nature follow their own respective trajectories), independent of human volition, control, manipulation, or intervention.

Ignoring the cosmological sense of nature entails that the question can only be raised and answered in any meaningful way by clarifying a distinction that is ontological in character (which is hinted at in the sense of nature as the contrast or foil of human artifacts), between nonhuman nature, on the one hand, and human culture, on the other. This distinction is not intended to be a dualism but a dyadism. Following Descartes, the former privileges one side and inferiorizes the other. For instance, mind is privileged over body, male over female, culture over nature. In contrast, in a dyadism, like the nature/culture dyadism advocated here, there is no master and no underclass.

According to the notion of dyadism, a distinction is retained not to embody the superiority of either culture over nature (as in anthropocentric thought) or of nature over culture (as nonanthropocentrism is sometimes accused of) but simply to mark the ontological difference between what is human (culture) and what is nonhuman (nature).[6] Humans, given their unique consciousness, their ability to think abstractly, and their more and more powerful technologies, which enable them systematically to transform nature to suit their own ends, have set themselves apart from nature. Human consciousness is capable of recognizing that nature, its processes, and its products belong to a different ontological category from that of humankind itself, its intentional activities, and the products generated by such activities. The difference is ontological in character because while cultural products, ex hypothesi, exist only as the result of human intentions, activity, and manipulation, nonhuman natural processes and their products (in principle) come into existence, continue to exist, and go out of existence entirely independently of human control and manipulation.

A world without human beings is necessarily a world without human cultural products; when the last human has expired, the Taj Mahal no longer exists as the cultural artifact we recognize it to be today—namely, as a monument erected by Shah Jehan to the memory of his beloved wife that we regard as aesthetically so beautiful and culturally so important that we have designated it as a World Heritage Site—but merely as a pile of stones. Such a realization in turn leads to the further recognition that there may be something morally amiss about the totalizing project of humanizing nature, entailing the ultimate elimination of nature as the ontological other.[7]

The scope of reference assumed in this discussion, as now defined, concerns all the other six senses identified above, although it highlights the sense of nature as contrasted with human artifacts. Regarding nature as pristine nature, some commentators have said that it has no application, as every part of Earth has been walked or worked on by humans. This assertion may be contested, however, as not all parts have borne human impact, as already observed earlier. But this sense needs to be discussed in conjunction with the sense of nature as that which bears or has borne human impact, distinguishing between different types and degrees of impact that have already been alluded to; treading on newly fallen snow (on Earth) or taking a few steps on the moon is very different, of course, from draining a regional swamp or building a vast dam.[8]

In general, the more permanent or wide-ranging the consequences of the impact, the more they undermine the autonomous status of that part of nature, and the more such affected nature becomes assimilated to human culture. Where nature has been but lightly impacted upon by humankind, such as when a faint trail has been created or an elementary shelter put up, one may still say that it is autonomous because its existence, which owes nothing to us, has not been made over systematically by us humans to suit our own ends, so that it can continue to exist, to flourish or not, as the case may be, independently of humankind. But where it has received maximum impact, such as when virgin rainforests have been turned into rubber plantations, then that part of nature has been systematically transformed to become a human artifact.[9] Antarctica, as such, may be said to be more or less pristine, given the light human impact it has so far borne. As yet no mining is permitted, and only some scientific bases run by various sovereign states are allowed, while in the recent past a few whaling stations dotted here and there were built to serve the whaling industry.

Nature as natural kinds (sense 4) and nature as elements in the periodic table (sense 5) may be briefly commented on jointly. According to today's scientific consensus, elements like carbon and hydrogen came into existence at the beginning of the universe as a result of the big bang. Atoms of carbon, oxygen, or hydrogen arrange themselves to form molecules; their peculiar arrangements give rise to the various familiar kinds like the different kinds of rocks, minerals, plants, and animals. Historically, nature in both senses long antedates the appearance of humans. Until the last seventy years or so, humans had neither the scientific understanding nor the technology to create any of the elements in the periodic table. But to mention a spectacular example, some physicists succeeded for the first time in manufacturing plutonium in the 1940s, an element that does not naturally occur and so is an artifact. Others have subjected a form of carbon to great pressure and produced artificial diamonds as a substitute for the diamond that exists as a natural kind. Of biotic kinds, humans have long learnt to domesticate a very small number of plants and animals; the vast majority of biotic kinds remain natural, but their future is extremely worrying given their rate of extinction today.[10]

To sum up briefly, the relationship between some of the various senses—nature as pristine nature (sense 2), nature as natural kinds (sense 4), and nature as elements in the periodic table (sense 5)—are all parts of nonhuman nature (sense 6). Nonhuman nature itself is part of nature in the cosmological sense (sense 1), but it is the part that excludes humans; the concepts of human and nonhuman are both subsumed under the cosmological sense of nature. Nature as the ontological contrast or foil to humans and their artifacts (sense 7) is concerned with that part of nonhuman nature that is primarily covered by pristine nature and partially covered by nature that bears or has borne human impact (sense 3), as clarified earlier above. As such, it has nothing to do with human intentionality and design, whether in its coming into existence, its continuing to exist, or the maintaining of its own existence. It stands paradigmatically for what could be meant by saying that nature is autonomous.[11]

Defining Autonomy

In what sense of "autonomous" may nature, in the way just identified, be said to be so? One needs immediately to get a red herring out of the way. This sense of autonomy should not be confused with the Kantian sense, in the domain of human action together with its related notions of freedom and reason as well as personhood, where Kant claimed that the categorical imperative holds sway: an individual will in legislating moral norms for itself is at the same time legislating for all free and rational fellow human beings.[12] Human consciousness and human intentionality render moral choice possible and the Kantian notion of autonomy appropriate.[13] The last section already touched on the sense of autonomy presupposed by the title of this contribution. It involves primarily the sense of nature as the ontological contrast to human artifacts (sense 7). This sense paradigmatically stands for what could be meant by saying that nature is autonomous. As the point is crucial, it is worth laboring and repeating what is meant by saying that nature is autonomous. It means no more and no less than the following: it simply is *what has come into existence, continues to exist, and finally, disintegrates/decays, thereby going out of existence, in principle, entirely independent of human volition or intentionality, of human control, manipulation. or intervention.* Its existence is independent not only of humankind but also of gods/supernatural entities. It is self-sustaining and self-generating.[14]

Nature, as identified and clarified in each of the various relevant senses above, is always autonomous in precisely this seventh way. This sense of autonomy has nothing to do with consciousness at all, never mind with reason and freedom. Parts of nature have no consciousness whatsoever and, of those parts that do (such as the higher animals), the consciousness displayed is not comparable to the human variety, which is unique.[15] For instance, Earth's mantle (in general), its core, its mountains, and the processes that brought them into existence and maintain them are autonomous, that is, entirely independent of humans. So is Mars—of course, humans hope, one day, to be able to terraform Mars and other heavenly bodies, but that day has not yet come. On Earth, humans may have flattened hills but not yet mountain ranges; neither do they yet know how to harness the natural processes behind plate tectonics to make new ones, even should they so desire.

History of the Universe

The history of the universe, as science understands it today, bears out the appropriateness of the notion of autonomous nature just identified; it shows that the term, at least, has reference. A quick summary suffices. The big bang is supposed to have occurred 4.5 billion years ago. Nature, insofar it stands for the elements in the periodic table, appeared; galaxies and all the bodies in them were formed from it. On Earth, the first form of life, prokaryotes, emerged 3.5 billion years ago; the eukaryotes (cells with a nucleus) arrived some 2 billion years after. Within several 100 millions of years following that event, four new kingdoms evolved: the protists (one-celled eukaryotic animals and plants), fungi, plants, and animals. Of the latter, in the Cambrian period, there evolved at least fifty different phyla, most of which became extinct. But the chordates survived, from which, eventually, mammalian life descended. Diverse vertebrates (which includes fish, birds, and reptiles) enjoyed a long period of ascendancy in the Paleozoic period before mammals became dominant. Among mammals, it was from the anthropoid apes, which are primates, that *Homo sapiens* finally emerged. If the age of Earth is made equivalent to a calendar year, *Homo sapiens* only appears three-and-one-half minutes before the year's end.

In virtue of the kind of consciousness unique to us,[16] via our science and our technology, humankind has come to uncover the sequence of phenomena just outlined above. The knowledge thus far attained about the past does not imply, however, that phenomena that occurred in the past, especially those before the emergence of *Homo sapiens,* are not autonomous in the very sense already identified. They were or are undeniably autonomous, as they came into existence, existed, and went out of existence in the absence of humans and were or are entirely independent of human intentionality and agency as well as self-sustaining and self-generating.

Nature in the senses picked out as relevant to the discussion of this paper is clearly not the product of human consciousness. The contrary claim, that nature is a mere social construct, is very problematic for my thesis. This paper must therefore argue against it, but, at this point of the presentation, it recognizes the claim only in the trivial sense that in the absence of a being with capabilities of conceptual understanding and linguistic articulation such as those possessed by humans, such a claim about it is impossible.[17] Trees, dinosaurs, or chimpanzees could not know the history of the universe as humankind has come to know it today, but they could *exist* prior to human consciousness.

Of course, humankind, as we have seen, has most certainly transformed nearly all of Earth's land surface in accordance with its intentions and purposes; houses, highways, theme parks, and mines are all human artifacts, objects that have been created deliberately by humans to serve specific human ends. The antonym of what is artifactual, however, is what is natural, that sense of nature as the ontological contrast to human artifacts that includes not only naturally occurring entities but also the naturally occurring processes behind physical, chemical, and biological phenomena and the interactions among them. The greenhouse effect as understood today is undoubtedly, in the main, anthropogenically caused; nevertheless, the activities and their going-on that constitute this effect are not themselves created and maintained by direct human manipulation. Furthermore, unlike the proposed terraformation of Mars or Earth's moon, the greenhouse effect has not been deliberately created by humans in the first instance (unlike, say, dams), although, in the light of the information and knowledge available to us through our scientific investigation of the matter, the continuing production of the greenhouse effect is the result of deliberate human intention.

Trajectory

Every naturally occurring entity or process has its own trajectory. The term "trajectory" is introduced to do the precise job of referring to the history as well as the character of each autonomous existence, whatever the entity or process may be. For example, a lake has its own trajectory. As a geological form, it is considered to be one of the most transient: it may dry out in a relatively short span of time, by becoming first a swamp and then, probably, a meadow. In the case of naturally occurring processes, these may be biotic, abiotic, or an interaction involving both. It is also the case that a species (for instance, *canis lupus*), as much as an individual member of the species (the particular wolf roaming at a particular time in a particular forest), may be said to have its own respective trajectories.

One should immediately point out that the trajectory of the species is not necessarily identical to that of any of its individual members.[18]The use of the term to cover both the biotic and abiotic domains does not entail, however, that their trajectories are identical, that what is true of the one is also true of the other. I have already noted that, as far as present evidence goes, the history of Earth shows that the abiotic long preceded the biotic on the planet. Present evidence also shows that without the continuing existence of a certain combination of abiotic conditions the biotic would not, and could not, continue to exist.

Two further points need to be emphasized. One concerns the crucial differences between the biotic and the abiotic; the other is to argue that, in spite of these differences, it is appropriate to use the notion of trajectory to talk about the entities and the processes in both domains.

First, the crucial differences. In general, individual organisms go through certain recognizable stages from their beginning to their end: infancy, growth, maturity, senescence, and, finally, death. They also possess certain characteristics, depending on the particular stage of their existence or, indeed, of their sex. For instance, in the trajectory of a frog, it starts off life as an embryo, which soon develops into a tadpole and then an adult frog. The tadpole clearly looks very different from the adult frog, yet the former is but a stage in the growth of the latter. Similarly, the peahen looks very different from the peacock, yet both are members of the same species.

By contrast, the mountain of granite remains a mountain of granite. Of course, even granite wears away over a large expanse of geological time. But granite does not mature to become something else: it might weather away to become soil, but it does not grow or develop to become soil in the same way that the tadpole becomes the adult frog. Granite may end up as soil, but soil is not granite: they are two very different things. In the case of the frog, however, the tadpole and the adult frog belong to the same species and are simply different stages in the trajectory of individuals belonging to it.

Another obviously important difference between the biotic and the abiotic is that the former appears to be an exception to the laws of thermodynamics, whereas the latter is not. But, of course, the appearance is only misleading. What misleads one is that life manifests certain processes at work that are absent in the case of the abiotic. Individual organisms are autopoietic; they maintain their own functioning integrity because they engage in metabolical and physiological activities. They therefore appear to produce order out of chaos, so to speak.[19] Again, by contrast, abiotic entities are not autopoietic or capable of reproduction, as they do not possess any mechanisms analogous to those found in organisms.[20]

In spite of the admitted differences, however, one may argue that the term "trajectory" may meaningfully be applied to both. Only the biotic may be said to be autopoietic, yet both biotic and abiotic nature may be said to be self-sustaining and self-generating in the larger senses of these terms.[21] I shall concentrate first on defending the assertion with regard to abiotic nature. Here, one must be a bit nuanced and distinguish among different sorts of abiotic nature and abiotic processes at work. For instance, mountains are self-generating (geologists call nature's "making" of mountains "orogeny") in the sense that certain geological processes, such as particular movements of tectonic plates, throw up mountains, like the Himalayan mountain range. But it is true that mountains are not self-sustaining; they wear down over the millennia through purely physical/chemical processes, which, however, are themselves self-generating and self-sustaining. On the other hand, other abiotic phenomena are both self-sustaining and self-generating. For instance, if the Gaia hypothesis is correct (and increasingly philosophers and scientists are taking it seriously), the maintenance of Earth's atmosphere is both self-generating and self-sustaining. And in other cases, the interactive processes between biotic and abiotic nature are both self-generating and self-sustaining.

But, to be more precise about the autonomy of the different trajectories in general of biotic and abiotic nature, one needs to draw attention to the important distinction between saying that an entity exists by itself, on the one hand, and that it exists for itself, on the other.[22] I argue that all naturally occurring entities, whether biotic or abiotic, exist by themselves; however, only the former exist for themselves.[23]

As I have already observed, naturally occurring entities and processes are precisely those that have come into existence, continue to exist, and go out of existence entirely autonomously and therefore independently of human intentionality and agency (and of supernatural agency, for that matter). They do not owe their being in any way to humankind. They are also self-generating and self-sustaining. So, insofar as they exist, they can then be said to exist by themselves. Naturally occurring biotic entities, however, display an additional dimension of complexity in their existence, as I have also already observed. In existing by themselves, they also necessarily exist for themselves, given that they are members of organic species, unlike abiotic entities, which are members of inorganic natural kinds. As organisms, they strive to maintain their own functioning integrity.[24] They develop, grow, mature, replicate (and in the case of the higher animals, they nurture their young), and their success in these activities involves active appropriating suitable nutrients and protecting themselves against adverse circumstances that would either harm or kill them. In the case of the higher animals, there is no doubt that they perceive danger and take steps to avoid it.

But skeptical voices are raised, particularly with regard to plants. Lacking any obvious organs of perception, what steps can they take to avoid harm and danger? They may not be as inept as they are often made out to be, however. For instance, when there is a drought, plants, unlike animals, cannot uproot themselves from one location to move to another, it is true. But they can grow deeper roots so that they may tap water lying farther down in the soil or send signals to their stomata to close up somewhat so that their rate of transpiration is reduced. As scientists research more into plant behavior, they have found that plants are not helpless in the face of adversity but have numerous and various mechanisms to overcome adversity. In other words, their behavior is neither random nor haphazard but goal-oriented, in spite of the fact that they lack consciousness. So impressed are some biologists by this aspect of the behavior of all organisms, whether plants, animals, fungi, or bacteria, that they have coined the very term "autopoiesis" to characterize it.[25] To conclude, one could say that both biotic and abiotic natures, in their diverse forms, follow their respective self-generating and/or self-sustaining trajectories.

Different Theses of Teleology

To reinforce the distinctions and definitions outlined above, one also needs to distinguish among three different theses of teleology: external teleology, intrinsic/immanent teleology, and extrinsic/imposed teleology.

External teleology refers to the view best illustrated by the famous paragraph in Aristotle's *Politics* (1, 8, 1256b16–22) from which the medieval notion of the Great Chain of Being is derived. It refers to a hierarchy of being with humankind at the top, followed by animals and then plants (and, presumably, then abiotic entities). Plants exist to serve the ends of animals, and animals in turn exist to serve the ends of humans. This thesis of teleology has lost its potency as Darwin's theory of evolution through natural selection has come to establish itself within the scientific community as the paradigm. In the light of Darwinian understanding, it is obvious that plants did not come into existence to serve the ends of either animals or humans; neither is it true that animals came into existence to serve the ends of humans. In the history of evolution, both plants and animals came into existence autonomously. They came into existence and continue to exist solely by themselves, their respective trajectories being entirely independent of humankind. In other words, the thesis of external teleology is simply false.

As already noted, individual organisms, unlike abiotic entities, in existing by themselves also exist for themselves. This follows as an implication of denying the thesis of external teleology. As autopoietic beings, they strive to keep alive, to reproduce, and so on, not to fulfill any end or purpose of any external agents but entirely to maintain their own functioning integrity. They do what they do for themselves alone.[26] In appropriating nutrients to sustain itself, the oak produces acorns in order to reproduce itself; it does not produce the acorn in order that the pig may have food. In turn, the pig eats the acorn simply to maintain itself and not in order to provide a nice dinner for the human hunter. And when the lion eats the human who has just dined off the roast pig, it is just as true to say that the human has not eaten the pig in order that he might himself in turn satisfy the lion's hunger. Of course, as a matter of contingency, organisms in general find certain other organisms useful in sustaining their own functioning integrity; plants that are primary producers still find insects helpful in propagating their pollen and certain mammals useful in propagating their seeds.

Organisms, in living for themselves, are realizing their respective tele as individuals and as members of their species.[27] In so doing, they exemplify the notion of intrinsic/immanent teleology. An adult female frog will mate with its male counterpart. The embryos she produces will develop into tadpoles; in turn, the tadpoles will grow into adult frogs. As they are frogs, not birds or wolves, they can only live or thrive in certain habitats; they prey on some organisms, like insects, but not others. In all ways, they behave as they do entirely in accordance with their own tele, independently of human agency and its manipulation. Their trajectories, both as individuals and as species, have, in principle, nothing to do with humankind. In the absence of humans in the world, they would be there, coming and going, at their own pace and in their own ways.

Humans, however, may turn naturally occurring organisms into artifacts, just as much as they turn abiotic matter into artifacts.[28] It may be helpful to recall the short definition of "artifact" already given earlier: it is the material embodiment of human intentionality. As such, it exemplifies the notion of extrinsic/imposed teleology. Humans have been domesticating plants and animals for a very long time. For millennia, their success rested on using what today is called the craft technology of selective breeding. But in the first half of the twentieth century, these traditional methods were radically overhauled by a new technology that was informed by the theoretical understanding given by the science of classical Mendelian genetics. The last quarter of that century also saw the arrival of a yet more powerful technology, sometimes called biotechnology or genetic engineering, that is informed by the theoretical understanding given by the even more basic sciences of molecular genetics and molecular biology. It is more powerful precisely because it allows humankind to cross boundaries between species and kingdoms by manipulating organisms, no longer at the level of whole organisms but at the molecular—DNA—level. For instance, one can insert into bacteria DNA that may belong to the human genome, or one can get cows to produce human proteins in their milk.

The examples just mentioned illustrate the process of transforming naturally occurring organisms, in the case of the bacteria, to become biotic artifacts and, in the case of the cow, which as a domesticated animal is already a biotic artifact, to embody a deeper level of artifacticity. The transgenic cow, unlike the more usual domesticated cow, has been commandeered by humans to use its autopoietic powers of self-maintenance to produce not cow's milk but milk that contains a human protein. In other words, biotechnology has succeeded in severing, in the clearest manner possible, what has been an inseparable link between being an organism that exists by itself and one that exists for itself. Up to even twenty-five years ago, the distinction between "by itself" and "for itself" was could only be made intellectually, not empirically. But recently biotechnology has managed to sunder them as a matter of fact.

The transgenic cow, par excellence, no longer exemplifies existence by itself; *ex hypothesi*, such an organism would not have existed without the direct and deliberate intervention of humans. The same is true of the transgenic bacteria. Humankind, via biotechnology, has captured the biological mechanisms of cows and bacteria in order to make them be what humans want them to be and not how they themselves would be in the absence of human manipulation and control. In other words, although they may still perform such biological functions, nevertheless, in carrying them out, they have been made to subvert their own respective tele. The cow no longer produces cow's milk, fit to nourish her own offspring, in principle at least, when the milk is not whisked away for human consumption. Instead, she is made to produce milk whose constituents are not those in accordance with her telos as a cow. The same is true of the bacteria: their own tele have been subverted and made to execute a human intention and human end instead. This embodies the notion of extrinsic/imposed teleology.

As I have just shown, the fact that organisms, in maintaining their own functioning integrity, exist for themselves has not stood in the way of human success in transforming them to become biotic artifacts. Biotic artifacts, as much as abiotic artifacts, are not autonomous, as they are no longer *simpliciter* naturally occurring entities. In spite of the profound similarity of sharing the same ontological status of being artifactual entities, however, there is a residual difference between them. Imagine the sudden disappearance of *Homo sapiens* from the face of Earth. Empirically, over time, abiotic artifacts like houses, jewelry, and computers will cease to exist; such artifacts, without constant human maintenance and repair, will disintegrate and eventually become dust and/or soil. At the philosophical level, in the absence of humans and their type of consciousness, there would necessarily be no human artifacts, just chunks of physical matter, even before they disintegrate and decay. The other sentient beings that remain, like the leopard or even the chimpanzee, would and could not know these things as human artifacts; only another being with a consciousness similar to ours could recognize them as such.

In the case of transgenic cows or pigs (or indeed of the more ordinary nontransgenic ones), in the absence of human maintenance, many would die or not succeed in reproducing themselves. But some might survive, and after many generations over evolutionary time, the humanly selected characteristics or the inserted transgenic material might become very rare in the genetic makeup. They could become naturally occurring again, like feral pigs, except for their remote genetic history. What this implies simply is that natural evolution (in the absence of human manipulation and control) has its own trajectories.

Nature as a Social Construct

The claim that nature is autonomous, especially when nature is understood as the ontological contrast to human artifacts (sense 7), could appear to be undermined by the counterthesis that nature is a heavily contested site, that all accounts of nature are necessarily socially constructed, implying that all claims about nature—such as whether nature exists or whether it is autonomous—are simply *au fond* the creation of human consciousness or are about the human interests behind these various representations. Some arguments would be needed to meet the charge; however, given limitation of space, any discussion must necessarily be brief and cannot hope to deal with all the complexities of the debate.[29]

Those who advocate that nature is socially constructed assume it is clear what their thesis really amounts to. But is it? At least four very different interpretations may be distinguished:

1. Any account of nature is necessarily anthropogenic. This is, obviously, because only humans can delineate terms and define them in certain ways. Defining terms is a linguistic activity. Human consciousness is necessarily mediated via language, and language is public and social, not private and individual.
2. Any account is necessarily anthropocentric. As it is humans who delineate and define terms, these are bound to be filtered through the human perspective, thereby reflecting and embodying their interests, preferences, or biases.
3. Any account necessarily reflects, embodies, and therefore implicitly advances the interests, preferences, or biases of particular groups in society. It is a mistake to think that all humans constitute a single group politically, socially, across time and space. On the contrary, they are fragmented into different groups along the fault lines of gender, religion, race, class, and so on. Each group may construct its own account of nature; however, the dominant account is that of the most powerful group in social and political terms.
4. Any account, whether anthropocentric or nonanthropocentric, necessarily is advanced by a particular group or individual whose behavior in turn is necessarily socially and normatively constituted.

Interpretation 1 is obviously correct, but it is a truism and therefore innocuous. Human thought and (human) language are indeed inextricably intertwined. As such, language does lay down the limits of what can be said at any particular time. But so long as this is not taken to imply that what can be said exhausts what is (and what can be thought), it remains innocuous: language can be stretched via metaphors, the introduction of new terms minted from old or dreamed up afresh to enable new thoughts and insights to be articulated, discussed, and critically appraised. Nothing, of course, can be asserted or described without recourse to language, which humans alone possess.[30] To lament that this is so is pointless. To admit that any account or representation of nature (or whatever) is necessarily mediated by language is not to admit that such accounts or representations necessarily are simply self-contained linguistic constructs created by those who put them forward.[31] In the same vein, one may also grant the associated quasi-Kantian point that any account of the world is necessarily filtered through a set of conceptual lenses and that there may be more than one set possible.

The central claim of interpretation 2, namely, that any account of nature is necessarily anthropocentric, itself rests on the prior tenet that anthropocentrism is uncontested and correct. But this is not the case. Many environmental philosophers have seriously challenged this dominant worldview.

Unless anthropocentrism is assumed to be correct, the interpretation cannot be sustained.

To see why this is so, it must first be pointed out that interpretation 2 appears to fail to distinguish between anthropogenic and anthropocentric. An anthropocentric worldview is one that claims humans alone to be morally considerable or to have intrinsic value (because they have language, interests, preferences, or whatever) and that always prioritizes human over nonhuman interests. "Anthropogenic" simply means "caused/produced by humans." One must grasp that only humans can articulate worldviews, whether anthropocentric or nonanthropocentric; as such, worldviews are necessarily anthropogenic. "Anthropocentrism" is not the antonym of "anthropogenic"; that antonym is "nonanthropogenic," which means "not caused/produced/generated by humans."

The failure to grasp the distinction enables the thesis to pass as true in the following way: as humans alone are capable of articulating interests or preferences, the only interests or preferences they can articulate are those that directly concern themselves only. Thus, from an anthropogenic premise, an anthropocentric conclusion is derived. But this reasoning as it stands is unsound. The conclusion does not necessarily follow unless it already presupposes that only beings that are capable of articulating their interests via linguistic categories are morally considerable. This additional premise would then rule out nonhuman beings from the domain of moral considerability, as they necessarily cannot articulate their interests or preferences. It follows, too, that humans do not owe them direct duties to protect or safeguard their interests or preferences. But the additional premise and its implications are precisely those (as earlier mentioned) that nonanthropocentrists, of one kind or another, strenuously challenge.

It is not obvious, then, in the light of the clarification above, that all accounts of nature are necessarily socially constructed in the sense of necessarily advancing anthropocentric interests and preferences. Some are clearly nonanthropocentric. One may then conclude that interpretation 2 is either an innocuous truism—all accounts of nature are necessarily anthropogenic—or simply false, as not all accounts of nature are necessarily anthropocentric. Of course, the worldview of anthropocentrism itself is determined by certain social/psychological considerations on the part of those who advance it; however, there is no inevitability about it as it is obviously possible for humans to advance the opposite worldview, namely, nonanthropocentrism. Both, however, are necessarily anthropogenic.

Interpretation 3 appears to embody some valid insights and points toward certain legitimate issues: (a) Humankind is indeed splintered along certain fault lines. According to Marxist thought, the most important of these is class: the capitalist class will therefore have a view about nature that is different from that of those who do not own capital. According to the affluent, nature is a place for recreation, for recharging one's spiritual batteries; according to the poor, nature is a place where one struggles to keep alive, oblivious of the landscape, however beautiful or sublime. By and large, according to the affluent middle classes in the industrialized West, nature (as wilderness) is not a place to live but to visit; according to the dispossessed, nature as wilderness means physical and spiritual exile. (b) Behind the various contested accounts lies a question of social justice. For instance, some developing nations (or the political elites of some developing nations) are sometimes not so keen to refrain from development using technologies that may be very ecologically damaging, arguing that it is more important to raise their material standard of living using the technology they can afford. Tigers should be saved, but ought they to be saved at the expense of those people whose livelihoods may be destroyed or undermined if more space were given to tigers?

In this view, the plea to save nature is not to be understood as a direct moral defense of nature but as a plea on the part of a certain group to protect its own interests. Environmental justice is distinctly anthropocentric in orientation; it is always about justice among different human groups and not between human communities, on the hand, and nonhuman groups, on the other. While some people, however, may have great sympathy with the struggle for social justice in a world with extreme inequalities inter and intra economies, it remains true that not every account of nature is necessarily socially constructed in the sense that it is the result of differential economic and political power relations between groups. For instance, a nonanthropocentric account of nature in favor of saving biodiversity does not per se advance any sectional human interests, indeed any human interests at all, now or in the future, as its avowed aim is to protect certain values in nonhuman nature embodied in individual organisms, in species, or in populations of organisms within different ecosystems that could be jeopardized by human activity. One could say that its explicit goal is to articulate not so much human interests as the interests of plants and animals, either as individuals and/or as species.

Of course, this is not to deny that many supporters of biodiversity act out of anthropocentric motives, in that they are saving biodiversity not so much for themselves in their own life times as for human posterity in general. But to maintain a priori that such a goal necessarily is but a smokescreen for sectional human interests, or even for the interests of humankind in the future, is objectionable, as reducing all accounts of nature to sectional/nonsectional human interests runs into the danger of confusing the anthropogenic with the anthropocentric. As already observed, any worldview, whether anthropocentric or nonanthropocentric, requires humans to articulate it; all worldviews are necessarily anthropogenic. To insist, in the absence of empirical evidence, that a nonanthropocentric perspective is a mere disguise for advancing the interests (sectional or nonsectional) of its advocates is to turn the claim into a metaphysical one ("metaphysical" in the abusive sense of the term as "totally unfalsifiable"). It is analogous to the equally metaphysical claim that all humans are egoists even in the face of evidence that humans can, and do, act altruistically some, though not all, of the time.

Interpretation 4—taking Steven Vogel as the source of this quasi-Habermasian account—does not confuse anthropogenic with anthropocentric, as it accepts that humans alone are capable of articulating worldviews, which can be either nonanthropocentric or anthropocentric.[32] Nevertheless, it says that humans, in speaking on behalf of nature, are necessarily socially and normatively constituted. For instance, what scientists say nature is cannot be taken to be objective and value-free, as scientists themselves are socially and normatively constructed.

In order to determine whether this interpretation undermines the thesis that nature is autonomous, one needs to distinguish between two types of anthropocentrism, what may be called axiological anthropocentrism—that only humans are intrinsically valuable—and existential anthropocentrism—that humans, who alone can be participants in communication, as moral subjects, are necessarily socially and normatively constituted. The former is incompatible with a nonanthropocentric account of nature; the latter is perfectly compatible with such an account. But existential anthropocentrism, though undoubtedly true, is as innocuous as the thesis closely related to it, namely, that all accounts of nature are anthropogenic, provided it is not understood to mean that all knowledge is reducible to the sociology of knowledge, to embrace philosophical or normative relativism, or that reality or nature does not exist outside the activities of such socially and normatively constituted beings and their representations of reality or nature.[33] In other words, all accounts of nature are necessarily both anthropogenic as well as anthropocentric in the existential sense but not necessarily anthropocentric in the axiological sense used above. But Vogel does not endorse axiological anthropocentrism, only existential anthropocentrism.

To conclude: the thesis that nature is socially constructed turns out under critical scrutiny not to be a potent knock-down argument against the possibility that nature may be said to be autonomous. Interpretations 1, 2, and 4 lose the challenging bite they seem to have at first sight once certain distinctions have been made. Interpretation 3 does seem to pick up on some valid themes, which have to do with an important issue of political/social justice in an unequal world, rather than on anything potentially philosophically challenging to the thesis that nature is autonomous. It has merits provided it is not understood as holding the indefensible position that any championing of nature's autonomy or of nonanthropocentric concerns in general is merely and necessarily a disguise for championing anthropocentric ones, whether sectionally construed or not. If it were to do that, then it would have turned itself into a metaphysical (in the abusive sense of this term) claim.

Conclusion

It is not self-evidently true that nature is not autonomous. Posing the question itself—is nature autonomous?—is perfectly possible and intelligible. To justify answering it affirmatively requires some careful clarification of certain relevant concepts, such as those of nature and autonomy, and the introduction of a new notion like trajectory. It also involves distinguishing between saying that an entity exists by itself and saying that it exists for itself. While all naturally occurring entities exist in the former sense, only naturally occurring organisms exist both by themselves and for themselves. In living for themselves, such organisms also necessarily live by themselves. Technology, however, whether in the form of traditional breeding techniques or, increasingly today, in the form of genetic engineering or biotechnology, has transformed certain animals and plants into biotic artifacts, such that in living for themselves, they no longer necessarily also exist by themselves. As such, their tele have been compromised; they no longer live out their own tele but have become the embodiment of human intentions and human ends.

That is why it is so important to distinguish among three theses of teleology. External teleology is simply false in the case of both biotic and abiotic entities. Intrinsic/immanent teleology obtains in the case of naturally occurring organisms, while extrinsic/imposed teleology is true of all artifacts, whether biotic or abiotic. To say *tout court* that the end of nature is nigh is too simplistic. The surface of Earth has certainly been severely made over by humankind. But to acknowledge this truth does not necessarily entail that nature, in many ways and in many contexts, was and is not autonomous. Neither does the claim that nature is socially constructed turn out to be quite the threat to the view that nature may be said to be autonomous that it appears, at first sight, to pose.

NOTES

1. Bill McKibben, *The End of Nature* (New York: Random House, 1989).
2. A definition of "artifact" is given in n. 5.
3. On natural kinds, see T. E. Wilkerson, Natural Kinds (Aldershot: Avebury, 1995).

4. Of course, as far as we know at the moment, only Earth sustains the biotic. The rest of the universe may well be bereft of life, although the possibility of finding it on some other planet elsewhere, either in the past or in the present, cannot be ruled out a priori.
5. In the discussion that follows, any entity or process that answers to this sense in particular is said to be naturally occurring. On the other hand, an artifact may very briefly be defined as the material embodiment of human intentionality.
6. I do not wish to deny that animals have culture and that some animals do create their own artifacts. For example, the beaver makes its own dams. The dyadism as understood in this context is required for the purpose of discussing whether we humans, with our peculiar type of consciousness, have a moral duty to nonhuman nature, including those animals that have their own culture, create their own artifacts, and have come into existence, continue to exist, and will go out of existence independently of human existence and human intentionality.
7. For further clarification of the ontological character of the distinction, see Keekok Lee, *The Natural and the Artefactual: The Implications of Deep Science and Deep Technology for Environmental Philosophy* (Lanham, Md.: Lexington Books, 1999), 180–84.
8. Every human activity, no matter how seemingly trivial and insignificant, such as walking lightly on fallen snow, is undertaken within a cultural context. Some could be treading lightly on the snow because they like to be in the great silent, white outdoors; others could be doing the same because their religion bids them to visit a shrine in the forest and to pray there specially after a snowfall.
9. The rubber trees planted in rubber plantations are not wild but domesticated plants. Domesticated plants are biotic artifacts, though their degree of artifacticity is not as deep as the transgenic tomato plant. In the wild, rubber trees do not in any case grow in plantations, as monocultures covering several hundreds or thousands of hectares. Instead, they are found scattered here and there in the forest.
10. Of late, biotechnology, consisting of techniques such as DNA engineering, has enabled humans to manipulate organisms at a newer and deeper level. But this point will be taken up later in greater detail.
11. See Ned Hettinger, "Respecting Nature's Autonomy in Relationship with Humanity," in this volume, for a critique that it is not enough to concentrate on such a sense for an adequate environmental philosophy; see also "The Autonomy of Ecosystems," section 3 of William Throop and Beth Vickers, "Autonomy and Agriculture," in this volume, which also argues that this sense is too strong.
12. Because of limitations of space, I will not comment on the concept of autonomy understood in today's moral philosophy.
13. Of course, strict determinism denies it.
14. In a later section, these characteristics will be further discussed within the context of the notion of the trajectory or trajectories.
15. In this volume, see Hettinger as well as section 3 of Throop and Vickers for clarification regarding this point.

16. This appears to be the case as far as evidence thus far has shown.
17. These are just a few holding remarks; a much more detailed defense against the charge of social construction will be explored in a later section.
18. Suffice it to mention only one difference in their respective trajectories: the species will probably live for about a million years before becoming extinct, this being the average life span of a mammalian species, whereas the life span of a wolf is only a few decades. Note, too, that this view presupposes that species are real entities, a thesis that is contested; however, this is not the place to defend it against its critics.
19. This does not prove that they are a genuine exception to the laws of thermodynamics, however: to create order in their own structure and functioning, they have to absorb other ordered matter in the form of nutrients. They also produce disorder when they respire and excrete, and when they die, they disintegrate and decay.
20. This notable characteristic has led environmental philosophers to hold two opposing views regarding the biotic and the abiotic. The first maintains that the abiotic has no intrinsic value; only entities with interests can be the loci of intrinsic value. Rocks and rivers per se do not have interests in the way plants and animals do. Yet other theorists argue that they do have interests and therefore may be said to be the loci of intrinsic value; see, for example, Val Plumwood, *Feminism and the Mastery of Nature* (London: Routledge, 1993). I reject both of these views; for details, see Lee, *The Natural and the Artefactual.*
21. In other words, autopoiesis is a subcategory of the larger class of self-generating and self-sustaining natural phenomena. Organisms alone are autopoietic, that is to say, self-generating and self-sustaining in ways (following trajectories) peculiar to themselves; they have physiological/metabolic, neurological, hormonal, reproductive, and other mechanisms by which their trajectories are regulated.
22. For further arguments on this point, see Lee, *The Natural and the Artefactual,* pp. 161–66, 227.
23. This use of the term has nothing to do with Sartre's use of it in his existentialist philosophy.
24. The striving referred to here need not involve consciousness, as in the cases of plants and the lower animals, although it may do so in the case of the higher animals.
25. See Humberto R. Maturana and Francisco J. Varela, *Autopoesis and Cognition: The Organization of the Living* (Dordrecht: D. Reidel, 1980). In the judgment of this author, however, Maturana and Varela have spoiled their case by arguing that organisms are machines, albeit "autopoietic machines." To say this amounts to a contradiction from an ontological perspective: machines are artifacts, the material embodiment of human intentionality, whereas naturally occurring organisms are autonomous and exist by themselves as well as for themselves. The topic of domesticated organisms and transgenic ones will be raised later.
26. No volition should be read into this locution in the case of plants or the lower animals, however.

27. In this context, telos or tele (in the plural) is used to refer to the developmental program that inheres in every individual organism as a naturally occurring being. For example, an acorn, in accordance with its telos, would become an oak sapling, which would grow eventually to be a mature oak tree, producing in turn its own acorns.
28. An extreme example of this is the transgenic organism. For details of the argument that biotechnology (genetic engineering) has created the transgenic organism as a paradigmatic biotic artifact, see Keekok Lee, *Philosophy and Revolutions in Genetics: Deep Science and Deep Technology* (London: Palgrave/Macmillan, 2002), chap. 5. For an opposing view regarding biotic artifacts, see Ned Hettinger's contribution in this volume.
29. For a more detailed examination, see Lee, *The Natural and the Artefactual*, appendix 2. For another critique, somewhat different from this one, see Holmes Rolston, "Nature for Real: Is Nature a Social Construct?" in *Philosophy of the Environment*, ed. T.D.J. Chappell (Edinburgh: Edinburgh University Press, 1997). See also Michael E. Soulé, "The Social Siege of Nature," in *Reinventing Nature? Response to Postmodern Deconstruction*, ed. Michael E. Soulé and Gary Lease (Washington, D.C.: Island, 1995; and John Searle, *The Construction of Social Reality* (London: Penguin, Allen Lane, 1995).
30. One has in mind here the kind of language that permits novels, history, and theories to be written or formulated.
31. In any case, a good many theorists who call themselves social constructivists would not wish to uphold this view; see note 32 for details.
32. Steven Vogel, *Against Nature: The Concept of Nature in Critical Theory* (Albany: State University of New York Press, 1996).
33. Philosophical relativism holds that all knowledge claims are equally valid, as the criteria for determining their validity are internal to the system of beliefs of which the claims are a part. Normative relativism is a subvariety of philosophical relativism; its domain is confined only to moral claims or beliefs. Many social constructivists, however, distance themselves explicitly from these positions. See, for instance, James Proctor and Richard White in *Uncommon Ground: Toward Reinventing Nature*, ed. William Cronon (New York: Norton, 1995.

    Equally, many social constructivists also are happy to acknowledge that "nature is real," that "there is more to the world than just words," although we need words to talk about the world, or that the natural world is more than just "our representations of it" (Cronon, *Uncommon Ground*, pp. 457–58.).

PART II

Autonomous Nature and Human Interests

ARE THEY COMPATIBLE?

FOUR

The Liberation of Humanity and Nature

ERIC KATZ

IN *COUNTERREVOLUTION and Revolt*, Herbert Marcuse declared that "nature, too, awaits the revolution!"[1] Nature, in other words, has a possible future free of human domination. Without going into a detailed exegesis of the work of Marcuse or other critical theorists—of which I am not an expert—I would like to consider the meaning of this idea: that nature itself is open to a revolution, a liberation, a release from human domination. And, unlike Marcuse, I will examine this idea by the consideration of two concrete examples of the ethics of environmental policy.

Mainstream environmental ethics has perhaps been slow to adopt the ideas of domination and liberation as descriptions of the human relationship with the natural world, despite the widespread use of these ideas in critical theory. William Leiss's *The Domination of Nature* was, after all, a study of Francis Bacon and not a treatise on environmental ethics.[2] A notable exception, of course, has been the work of ecofeminist philosophers. At least since 1980, when Carolyn Merchant published *The Death of Nature*, ecofeminist philosophy has emphasized as its primary theme the connection between the domination of women and the domination of nature.[3]

But as early as 1977 John Rodman, with perhaps an ironic eye towards Marcuse's essay, published "The Libcration of Nature?" (with a question mark!), a classic critique of both Peter Singer's idea of animal liberation and Christopher Stone's proposal for the legal rights for nature as models of a new environmental consciousness.[4] If nature were to be truly liberated, Rodman argued, we would have to do better than extending utilitarianism to the animal kingdom or granting rights as convenient legal fictions to nonhuman natural objects. Taking as his symbolic act of defiance the freeing of captive dolphins, Rodman argued that we must resist the technological monoculture that is rapidly enveloping the contemporary world.

In my own work,[5] I have used the idea of domination and the idea that I take to be its opposite, autonomy, as critical markers in any analysis of the ethics of environmental policy. But my use of these notions has been fairly uncritical; I have always been reluctant to get into any serious metaphysical debates about the meaning of human nature or the nature of nature itself. I believe that such deep philosophical analyses and debates may impede the timely development of urgently needed justifiable environmental policies. Nevertheless, when I claim that nature should be treated as analogous to a human subject or (in other words) when I claim that preserving the autonomy of natural processes should be the preeminent goal of human activity regarding the natural world, I open the door for critical questioning about the metaphysical foundations of my position. I need to defend, at the very least, my lack of concern for studying the metaphysics of nature.

The precise locus of my problem concerns the existence and description of nature in itself, the nature of nature. If, as I claim, the autonomy and self-development of nature are to be respected, if, in other words, nature is to be treated as a moral subject, then we need some sense of what nature is in itself, outside the domain of human activity. But nature is only known through human activity, and, even more problematic, nature is continually modified by human activity. Thus both epistemologically and ontologically, nature in itself is "our" nature, the nature constructed by human thought and praxis. Now, according to Steven Vogel in his book *Against Nature*, the problem of nature in itself is also the crucial problem for critical theorists such as Marcuse and Habermas: "how to reconcile an account of knowledge as active and social . . . with the 'materialist' commitment to a nature independent of the human."[6] But this problem is more than a problem for critical theory; it is a problem for all of environmental philosophy or at least all of environmental philosophy that deals seriously with a robust nonanthropocentrism. Any account of environmental ethics that extends moral consideration beyond the boundaries of the human species would seem to require some idea of what nature and natural entities are in themselves, free of human influence and control. Whether we talk of interests, rights, self-realization, or whatever, we need to know what is good for nature in itself.

The problem is that we know and understand nature through human categories. For example, we use human conceptions of good to evaluate the processes of nature, the flourishing of natural entities and systems. The human interest in nature is the factor that focuses our perceptions and understanding of the natural world. If nature is understood in this way, it does not appear that it could ever be free of human domination, for the basic domination is epistemological: nature is only known through human thought. For the operation of a nonanthropocentric environmental ethic, we seem to require an idea of a nature that is autonomous, a nature that is analogous to a human subject, so that we can preserve and promote the interests of this nature in itself. But to think of a free and autonomous nature, it seems, means that we must think of a nature that is completely free of human influence, to think of nature in itself, outside of all human categories of thought.

But can we know what nature is in itself? Given our post-Kantian understanding of human thought, it seems unreasonable to think that we can know nature *an sich*. But is knowledge of the noumenal world of nature really required for the development of a nonanthropocentric environmental ethics? Perhaps I have described the problem in the wrong way. Perhaps there is no real need for a metaphysical examination of nature as such. Here is where my pragmatism begins to kick into gear; here, I want to avoid metaphysical speculations; here, I am willing to make do with the concepts and practices that we have at our disposal as practical moral philosophers.

And so let me offer a tentative solution to the problem of nature in itself. Is there a Nature outside the knowledge and activity of human society that can be a subject unto itself? Is there a Nature that can be liberated from human domination? To answer this problem, let us compare the problem of the liberation of nature with the liberation of humans. Given the limitations of our epistemology, we do not really know what humans are in themselves either. The Kantian analysis of the knowledge of physical nature applies to humans in their physical being as well. I do not know other human beings, or even myself, outside of socially constructed categories. All my relationships with all individual human beings and all human groups and institutions are mediated by cultural constructs and social roles. And yet in my relationships with other humans and human institutions I can meaningfully strive to end oppression and domination, to aid other human beings in achieving liberation, freedom, and autonomy. I do not require an idea of a human being in itself for a meaningful liberatory praxis.

So what does liberation mean? It does not mean the elimination of all social constructs and categories. A human being does not become liberated when he or she transcends all social and cultural roles, duties, and obligations. Even if such a transcendence were possible—which it isn't—what could it possibly mean? A pure human essence existing outside of all human history, free of all the rules of human social life? The prehistorical natural or biological human? Although such an abstract ideal may have a place in the conceptual analysis of the meaning of human life, it surely plays no part in our daily practice of working toward the liberation of individual humans and human institutions.

Regarding the liberation of humans, then, my point is this: we do not need an idea of an ideal human nature in order to understand practices of liberation and domination that we encounter in the everyday world. There are, of course, difficult cases. As a parent, for example, I have long been fascinated by the boundaries of education, socialization, indoctrination, and oppression in my relationships with my growing children. But the existence of gray areas and marginal cases does not in the least prevent me from recognizing the real oppression of children by their parents, and my parenting, I hope, is always guided by both an understanding of the appropriate uses, abuses, and limitations of my authority and a rather nebulous idea of a maturing autonomous human being in contemporary culture, the characteristics I hope develop in my children. Similarly, in the broader social and political sphere, we do not require an idea of an ideal human nature in order to oppose (for example) slave-labor practices, various forms of racial, gender, and religious discrimination, economic injustice, and imperialism. Our social context informs our decisions. What we mean by human liberation is embedded within our social categories, which may, of course, change, as society itself becomes liberated. So human liberation is the development of specific positive freedom-and-life–enhancing roles, not the elimination of all social constraints, commitments, constructs, and categories. Although there will continue to be difficult cases, our ethics and our social praxis are enough. We need not turn to metaphysical speculation on the essence of humanity to give content to our activities regarding human liberation.

Why is it not the same for our relationship with Nature? Why do we need an idea of a nature in itself, outside of all human categories of knowledge and action, to give content to a robust nonanthropocentrism? Surely our practical activities in their interaction with nature are enough to provide us with a sense of what is right and wrong. Do I really need an idea of Nature in itself, the nature of nature, to know that clear-cutting a forest is a form of domination, an injury to the autonomous development of the forest ecosystem? Do I really need an idea of a nature unmediated by human categories of thought and action to know that damming a free-flowing river interferes with the free and spontaneous movement of natural processes? Without denying that there will be difficult cases, it seems clear that we know what is involved in the domination (and hence the liberation) of nature. Environmentalist practice informs our decisions; we have no need for metaphysical inquiries into the nature of nature as such.

So to return to Marcuse's claim: nature also awaits the revolution, its liberation. Can we give a concrete example of what this means? In my previous work, I have often discussed a specific example of *domination*—the redesign and restoration of damaged forest ecosystems—but what of *liberation*? What does the autonomy of nature look like? In a recent paper, I discussed the ethics of beach replenishment and preservation projects, and I believe that this is a good case to highlight my idea of the autonomy of nature.[7] I must confess that I have a personal interest in beach erosion, since I live a good part of the year on Fire Island, a barrier beach off the coast of Long Island in the Atlantic Ocean. Fire Island is an interesting case—or so I hope—because it is a hybrid environment. The island is thirty-two miles long and, at its thickest, about a half mile wide; it is, essentially, a long sandbar. Although there is no large-scale commercial development, some sections of the island are densely populated with individual homes on small lots (most less than a quarter acre). But most of the island remains undeveloped. There is a unique wilderness area in the central part of the island—the Sunken Forest—and the island is home to several threatened and endangered species of plants and birds. In 1964 the federal government purchased the island and made it part of the national seashore, equivalent to a national park.

As with all barrier beaches on the eastern coast of the United States, Fire Island suffers from erosion. Individual homes, recreational beaches, and the wilderness areas are threatened by the loss—the movement—of sand. Whether a policy of beach replenishment and preservation should be undertaken is a question that raises interesting issues in technology, economics, social justice, and environmental ethics. I address these questions in the paper I mentioned above; here, I am only concerned with the idea of the autonomy of nature. Can we look at the problem of beach erosion and the environmental policy of beach replenishment from the perspective of the liberation of nature?

Presumably, to liberate nature in this case, to permit the autonomy of natural processes, we would adopt a hands-off policy regarding beach erosion and replenishment. Rather than try to mold and manipulate the beach environment, we would simply leave it alone, thus permitting both the natural erosion (and sometimes the natural build-up) of sand to continue. But Fire Island is not a natural environment; it is a hybrid area of wilderness, relatively undisturbed beaches, and single-family homes. There are concrete and wooden walkways, extensive bulkheading, and numerous boat channels and harbors. It is as much a built and human environment as a natural or wild one. This makes the entire idea of the autonomy of natural processes rather suspect. Only if we were systematically to eliminate all human-built structures and modifications to the shoreline could we begin to approximate a natural environment. And in that case alone could the idea of the liberation of nature on the island make sense.

In the real world, of course, the systematic elimination of all human structures on the island is not going to happen. So let us simply undertake a philosophical thought experiment. Imagine an island identical to Fire Island—thirty-two miles long, central wilderness area, threatened and endangered species—but without a permanent human presence. No houses, no harbors, no boat channels, no sidewalks or roadways, no bulkheading. On this imaginary island, what would the liberation of Nature be like? Clearly, it would be the continuation of the freedom from human impacts. The autonomy of nature would be the unfolding of natural processes on the island—and the island's interactions with the ocean—without the interference of humans, without human development and alteration of the land. Nature would develop in its own way, not subject to the designs, plans, or projects of humanity. And to say that Nature would develop in its own way does not imply that Nature itself has a plan, a telos; we are simply eliminating the dominating tendencies of human plans, human intentionality and design.

This imaginary island thought experiment shows, I believe, that we do not need a positive conception of Nature as such to understand the idea of the liberation and autonomy of nature. We do not need to know a Nature outside all human categories. Indeed, the idea of nature that we have on my imaginary island is an idea constructed by our science; it is a nature that we understand through human categories. But this does not make it any less autonomous. As long as it is not being molded and transformed by human impacts, it is a free and liberated nature. It may not be free of human domination in a metaphysical or epistemological sense, but in the realm of pragmatic environmental policy it surely is.

Now, this argument obviously rests on a comparison and analogy between the liberation of nature and the liberation of humans. Let me anticipate and answer a possible objection to this argument based on an obvious disanalogy in the comparison: human liberation generally involves both the liberation of individuals and the liberation of groups, but in the liberation of nature it is difficult to understand the liberation of individual natural entities. Once we begin to think about the liberation of individual natural entities, we enter a territory of endless debates in the realm of environmental ethics over the specific relevant moral characteristics of individuals: is sentience, self-consciousness, or life itself the particular characteristic that confers the possibility of moral standing? I believe that my argument and analysis concerning the imaginary Fire Island enables us to sidestep these questions about the appropriate characteristic for moral standing; indeed, one of the strengths of this pragmatic approach is that it enables us to avoid these debates.

Consider the case of rocks.[8] Does it make sense to say that rocks can be liberated? From within the traditional framework of the field of environmental ethics as it has developed over the last thirty years, we would be perplexed by this question. Unless one wants to adopt a Whiteheadian perspective—or some other form of panpsychism—one is left with the obvious problem that rocks are not alive, not sentient, and not conscious of their own existence.

Rocks have no good of their own. So how can one say that rocks can gain liberation and exhibit autonomy?

One possible answer from within the traditional range of arguments would be to emphasize a holistic perspective on the ideas of domination, liberation, and autonomy. Unlike human liberation, which generally involves the creation of conditions for the autonomy of a human individual, one can say that the liberation of nature is a holistic liberation, in which one creates the conditions for the autonomy of natural systems. The ecosystem or bioregion is freed from the oppression of human design and modification. Under this interpretation, individual rocks—or trees or squirrels—are not themselves liberated. The system in which these natural individuals are embedded is liberated. The system acquires (or reacquires) its autonomy, outside the demands of human purpose and intentionality.

Although I am inclined to support this holistic response to the problem of the liberation of rocks and other individual natural entities—after all, much of my work in the field of environmental ethics has been the explanation and justification of a community-based ecological holism[9]—the position I have sketched out through the use of the imaginary Fire Island actually permits a more radical response. Yes, individual rocks can be liberated. In the Fire Island case, liberation was seen to be the freedom from human impacts. Why can we not conceive of individual rocks existing in such a state, unmodified by human activity? We can again compare two sets of comparable rocks, one subject to human modification (through drilling, hammering, or painting) and one simply left alone. Which set of rocks is free of human domination? The answer is obvious.

What is interesting about the case of rocks is that there is a real-life policy debate in the United States that focuses precisely on this issue. Rock climbers use metal bolts to enable them to climb a cliff. The metal bolts are hammered into the rock face, to be used as fixed anchors to attach ropes, nylon slings, and other safety devices. The bolts clearly alter the natural face of the rock wall. But even more problematic is the fact that often the fixed anchors are left in the cliff after the climb so that they are in place for future climbs. The policy debate in the United States concerns the use of metal bolts in national (and state) parks and designated wilderness areas, for the Wilderness Act prohibits the installation of any permanent human structure in wilderness areas. In some places, there are so many bolts in a cliff that a metal bolt ladder appears to have been constructed.[10]

Although the resolution of the policy debate in the United States will probably be dictated by the courts—wilderness preservationist groups are battling rock climbers with the National Park Service caught in the middle—the point of this case is that it shows that it does make sense to think of rocks—and other inanimate natural entities—as being potential candidates for liberation in the sense that I have been using: freedom from human impacts. We let things be; we leave nature alone; we let natural entities, processes, and systems develop without the modification and molding of human purposes and designs. Clearly, a rock cliff without metal bolts is different from one with bolts. The cliff without the bolts exhibits a kind of autonomy and freedom that has been denied the modified cliff.

Does this examination of the idea of the liberation of nature help us in the understanding and determination of environmental policy? Return again to Fire Island, the real island with complex interacting human and natural ecosystems. I live on Fire Island, and I need to know what kind of environmental policies will be morally justifiable there. My argument shows, I believe, that, even in the case of hybrid environments, we ought to lean toward leaving nature alone. In most cases, the absence of human domination will result in the liberation and autonomous development of nature. Where we have a choice, we should choose the least intrusive policy. On Fire Island, for example, if we wish to protect the recreational beaches, the wilderness areas, and the endangered species, we ought to preserve them by a process of sand replenishment and not build permanent structures such as rock jetties and sea walls. In the rock-climbing case, we have to regulate the use of fixed anchors that remain permanently in the rock face, at least in protected national parks and wilderness areas. Of course, this is not the place for a full-scale discussion of policy alternatives, for such a discussion would require a detailed description of the specific facts of the concrete situation. My philosophical point about the formation of policy is merely this: we can make decisions about the autonomy of nature and its relevance to environmental policy without plumbing the metaphysical depths of noumenal nature, Nature in itself.

Marcuse believed that, after the revolution, not only would nature be liberated but humanity would create a new nondominating science, founded on a new sensibility of passivity, receptiveness, and openness that would involve "the ability to see things in their own right, to experience the joy enclosed in them, the erotic energy of nature."[11] I do not know if any of this is possible. I do not believe that we can ever escape the historical conditions of our science, our economic development, and our ethical practice. Our understanding and evaluation of nature will always be contained and determined by our historically based human categories of thought. Yet we can still understand that some of our environmental policies involve the domination of nature. It is incumbent upon us to avoid and to minimize these oppressive practices as much as possible. Only then can we respect the autonomous development of nature as a subject in itself.[12]

NOTES

1. Herbert Marcuse, *Counterrevolution and Revolt* (Boston: Beacon, 1972), p. 74.
2. William Leiss, *The Domination of Nature* (Boston: Beacon, 1974).
3. Carolyn Merchant, *The Death of Nature: Women, Ecology, and the Scientific Revolution* (San Francisco: Harper and Row, 1980). For the most succinct statement of the connection between domination and ecofeminism, see Karen Warren, "The Power and Promise of Ecological Feminism," *Environmental Ethics* 12 (1990): 125–146.
4. John Rodman, "The Liberation of Nature?" *Inquiry* 20 (1977): 83–131. Rodman is critical of Peter Singer, *Animal Liberation* (New York: New York Review and Random House, 1975), and Christopher Stone, *Should Trees Have Standing? Towards Legal Rights for Natural Objects* (Los Altos: Walter Kaufmann, 1974).
5. Eric Katz, *Nature as Subject: Human Obligation and Natural Community* (Lanham: Rowman and Littlefield, 1997), pp. 93–146.
6. Steven Vogel, *Against Nature: The Concept of Nature in Critical Theory* (Albany: State University of New York Press, 1996), p. 141.
7. Eric Katz, "A Pragmatic Reconsideration of Anthropocentrism," *Environmental Ethics* 21 (1999): 377–390.
8. I would like to thank Holmes Rolston for bringing this problem to my attention.
9. See Katz, *Nature as Subject.*
10. The facts of this case were reported by National Public Radio. Mark Roberts, "Don't Bolt Me In," *NPR Weekend All Things Considered*, August 8, 1998. The problem arose at El Dorado Canyon State Park and the Lost Creek Wilderness in Colorado. My thanks to Andrew Light for bringing this case to my attention.
11. Marcuse, *Counterrevolution and Revolt*, p. 74.
12. This paper was presented as a talk at Mansfield College, Oxford University, June 28, 1999, at the Oxford Centre for Environment, Ethics and Society, as part of a joint meeting of the Society for Applied Philosophy and the International Society for Environmental Ethics. An earlier version of this paper was read at a meeting of the International Society for Environmental Ethics at the World Congress of Philosophy, Boston, Massachusetts, August 12, 1998. I would like to thank the participants in both sessions for helpful comments and criticisms.

FIVE

Respecting Nature's Autonomy in Relationship with Humanity

NED HETTINGER

PRESERVATIONIST ENVIRONMENTAL thought involves the following interrelated ideas. Nature's value is significantly a function of its degree of independence from humanity. Naturalness or wildness is what most centrally grounds nature's value. When considerably modified by humans, nature loses much of its value and even its essential character. A strong conceptual separation exists between humans and nature. Nature is to be understood in opposition to humanity; nature is the nonhuman. Wilderness is thus quintessential nature. Respect for nature most importantly involves preservation of wilderness areas, free from significant human influence.

In the context of today's massive and ongoing humanization of the planet, these ideas have much truth and power. With perhaps half the planet's surface significantly disturbed by humans and half of that human-dominated (Hannah et al. 1993), valuing nature for its remaining wildness, separating humans from nature, and preserving wilderness are essential if nature as an independent other is to continue to flourish on this planet. But, as important as they are, these preservationist ideas—left by themselves and unsupplemented—have a dark side.

Most troubling is that such views of the human relation to nature make it difficult to envision a positive role for humans in nature. As the antithesis of nature, humans necessarily degrade and destroy it. But an adequate environmental philosophy must explain how humans can be something other than an ugly scar or nasty stain on the natural world. Purely preservationist views also fail to provide guidance for how humans should treat the nature with which we must interact. Specifying how to value and respect nonwilderness lands (and less than fully wild animals and plants), including the rural, middle, or working landscape, is also a crucial task for environmental philosophy. At best, pure preservationists tell us to minimize our use of such lands and entities. At worst, preservationists see such lands (and the animals and plants on them) as human artifacts totally lacking natural value. But an adequate conception of humans' relation to nature must allow for the possibility of respectful *use* of nature. Unsupplemented preservationist views fail to account for how respect for nature can go hand in hand with human use of nature.

I believe that a particular concept of nature's autonomy provides an important resource for responding to these challenges. Preservationist intuitions need to be joined with the idea of respecting the autonomy of nature. A healthy respect for the wildness of nature that is significantly uninfluenced by humans combined with a respect for the autonomy of nature with which humans are involved provides a far more adequate and comprehensive ethic of respect for nature than does either ethic alone. This essay explores the concept of respect for nature's autonomy and relates it to preservationist intuitions.

Problems for Pure Preservationism

Numerous environmental philosophers, including some of the most influential, accept some version of these preservationist ideas and are vulnerable to the criticisms just mentioned.[1] For example, Paul Taylor's fine book *Respect for Nature* (1986) is concerned only with respecting wild nature and intentionally leaves out discussion of proper treatment of nature that has been heavily used in the fulfillment of human ends. This type of nature Taylor calls "the bioculture," and, in his account, it is "part of human civilization," not nature (310). Although Taylor believes developing an "ethics of the bioculture" is an important task, he thinks it is not part of environmental ethics proper. This is unfortunate, for an ethic of how humans should treat the nature with which they live and work is of crucial importance and a central (if too often forgotten) task of a philosophical account of the human relation to nature. Additionally, one of Taylor's fundamental rules for respecting nature is the duty of noninterference: "We must not try to manipulate, control, modify, or 'manage' natural ecosystems or otherwise intervene in their normal functioning" (175). We are "required to respect their wild freedom by letting them alone" (176). While the duty of noninterference in wild nature is crucially important, suggesting that any human modification of—or involvement with—nature of any sort violates a prima facie duty to nature makes a positive conception of the human relationship with nature difficult to conceive.[2]

Eric Katz's conception of the value of nature and our obligations to it also leaves little room for a positive account of humans' relationship with nature. His characterization of the human-nature relationship suggests that any human use of nature is abusive. Katz says, "When humans shape and manipulate the natural world to meet their own interests, to satisfy their desires, it is a form of anthropocentric domination, the oppression and denial of the autonomy of nature" (Katz 1997: xxiv). But humans, like other species, must influence the natural world. Human survival, much less human flourishing, requires this. Katz's language suggests that humans—by their very nature—dominate, oppress, and subvert the autonomy of nature. For Katz, even well-intentioned human involvement with nature—such as restoration of degraded nature—is oppressive. Katz writes, "The re-created natural environment that is the end result of a restoration project is nothing more than an artifact created for human use" (97), and although "these restored and redesigned natural areas will appear more or less natural . . . they will never be natural" (98). For Katz, then, the human stain on nature is so toxic that once nature is soiled, it has been spoiled forever; it will never return. Given Katz's account, it seems impossible to envision an environmentally just future in which humans live in the natural world in a morally appropriate way.

At times, Holmes Rolston's account of nature approaches the same set of problems. In a powerful response to J. Baird Callicott's (1991) critique of the idea of wilderness as unpeopled places and to Callicott's suggestion that human interaction with nature might benefit nature, Rolston says: "The fallacy is to think that a nature allegedly improved by humans is anymore real nature at all" (Rolston 1991:371). This borders on the claim that only pristine wilderness is real nature. Such a view leaves no place for humans in nature. Although Rolston sometimes writes about rural nature, he conceives of it as a "hybrid" between nature and culture (Rolston 1988:330), suggesting that real/pure nature has been degraded in rural landscapes. Furthermore, Rolston has a decidedly "tradeoff view" concerning interaction between humans and nature. Although in various places Rolston suggests humans might add to natural value, the dominant story is that human interaction with nature is a loss for nature. To flourish, human civilization must trade in natural values in the pursuit of cultural ones (Rolston, 1994:85–86). While there is much truth in this perspective, it is important to allow for types of human flourishing that need not compromise natural value.

If we accept the troubling idea that "nature can be fully itself and thus have full value only when left undisturbed by human beings" (Kane 1994:71), we are left with the unfortunate suggestion that—from the perspective of nature's value—a policy of human/nature apartheid would be best. In the context of today's harmful human transformation of the planet, apartheid *is* a major part of what is needed. Leaving much of nature on Earth alone is an absolutely central part of any adequate environmental ethic. But this is not all that is needed, and an environmental ethic that suggests nature necessarily loses or ceases to be nature in any significant interchange with humans makes the human presence on Earth a tragedy for earthen nature. Environmental philosophy must ultimately articulate a constructive human-nature relationship that allows us, as John Visvader says, to "imagine giving more to the world around us than the gift of our mere absence" (1996:18). The alternative of either minimizing human influence on nature (Katz's ideal of human/nature apartheid) or sacrificing natural value for human good (Rolston's tradeoff view) fails to provide for such a positive role.

For humans to have something other than a purely negative and harmful role with respect to nature, we must distinguish between human *involvement* with nature and human *domination* of nature. *Modification* and *alteration* of nature must be distinguished from *mastery* and *control* of nature. If we define human alteration of nature as ipso facto degradation, humans who want to be respectful of nature will not be allowed to interact with it at all. Activities such as birdwatching from a distance would seem the extent of allowable interaction. Relatedly, we need to explain how certain types of human uses of nature need not be abusive and how humans can use nature as a means without necessarily using it as a *mere* means. If our use of other humans need not be devoid of respect for them, one would expect that our use of nature need not be devoid of respect and concern for its flourishing. A positive vision for humanity's role in nature might involve a partnership relationship between humans and nature, where humans use nature respectfully while nature does not lose and perhaps even benefits from the interaction. A symbiotic, mutually beneficial relationship with nature is the ideal.[3] Such an ideal should supplement, not supplant, a preservationist ethic that requires significant separation.

Varieties of Nature's Autonomy

A particular way of understanding respect for nature's autonomy can provide a means to address these concerns. Autonomy is a form of independence that is distinct from absolute independence (i.e., maximum avoidance of influence). Respecting the autonomy of others does not mean avoiding interaction with or influence on them. What respect for autonomy requires is that one not dominate or control the other. Thus nature's autonomy need not be compromised by human involvement as long as this involvement is not domineering, just as a person's autonomy need not be compromised by the involvement of others so long as they avoid dominating and controlling that person. Jack Turner puts a related point this way: "Although autonomy is often confused with radical separation and complete independence, the autonomy of systems (and I would argue, human freedom) is strengthened by interconnectedness, elaborate iteration, and feedback—that is, influence" (1996:113).

Nature is clearly not autonomous in some senses in which persons are autonomous. With the possible exception of psychologically sophisticated animals, natural entities or systems do not survey the range of possible alternatives and intentionally choose a plan of action. Neither is the activity of natural entities autonomous in a sense that would justify us holding them morally accountable for their behavior. Nevertheless, the behavior of natural entities can be plausibly described as autonomous in a number of respects. Human action can be seen as thwarting or respecting nature's autonomy in these senses.

The most fundamental sense of nature's autonomy is freedom from human domination and control. Call this the autonomy of nature in relation with humanity. This is a purely negative sense of autonomy, and it consists in the lack of a certain type of human influence over nature. Autonomy in this sense is a relational property between natural entities and humans. To say nature is autonomous in relation to humanity is to say that nature carries on independently of human control or domination. Humans dominate nature when they exercise mastery over it by exerting the supreme determining or guiding influence. When human influence over a natural entity or process outweighs all the other determining factors, humans are dominating that entity or process and failing to respect nature's autonomy in relation with humanity. For example, regulating the height and timing of a geyser by the systematic application of soap in the underground plumbing undermines the geyser's autonomy, for humans are then the chief determining factor in the geyser's eruptions. In contrast, watering a tree so it grows more quickly and larger (or shading it with one's house and thus slowing its growth) does not dominate the tree, despite constituting significant influence over it.

To respect nature's autonomy in relation to humanity is not to respect nature in virtue of particular properties it possesses but to treat it in a certain way. We respect nature's autonomy by avoiding exerting the preponderance of influence over it. Humans can respect the autonomy of nature in this sense whether the natural entity is goal-directed (e.g., organisms and perhaps some ecosystems) or not (e.g., rocks and mountains), whether the natural entity is quite active (like a river) or relatively passive (like a pond). A natural arch about to collapse because of anthropogenic acid rain has had its autonomy undermined as much as has a drive-though sequoia whose life cycle has been cut in half by the tunnel, even though the former is not a self-organizing or teleological being. In both cases, humans dominate these natural entities by exerting the preponderance of influence over their fates. Similarly, keeping an arch from falling as a result of wind and water erosion by using metal cables and bolts puts humans in control of the fate of the arch and fails to respect its autonomy from humanity.

All natural entities and processes have headings or trajectories in the minimal sense that they have beginnings, endings, and patterns of change.[4] Humans can participate in and influence these journeys while respecting the autonomy of the entities undergoing them by avoiding a controlling or dominating influence. For example, human mimicking of the natural fire regime in a fire-adapted forest is significant human involvement in a natural system that nevertheless does not constitute domination or mastery over it in part because the overall trajectory of the system is not altered. Selective, multiage logging that preserves forest species and successional processes might also be compatible with the forest's autonomy from humanity for similar reasons. Affecting the population of deer and the predators that feed on them by subsistence hunting influences this predator/prey system without necessarily controlling it. In contrast, regulating deer and predator numbers by scientifically managed hunting seasons, a birth-control regime for the deer, or systematic poisoning of predators approaches human mastery over this predator/prey relationship and thus fails to respect the autonomy of this natural process.

It is sometimes suggested that if humans are necessary conditions for the existence of an entity, then it is ontologically dependent on them and thus lacks autonomy in relation to humanity (Katz 1997; Lee 1999). Domesticated animals and plants would not exist but for humans and thus, the argument goes, are dominated and controlled by humans. Insofar as this is a critique of contemporary agriculture, it is much too broad. According to this account, all agriculture, whether it be small-scale organic farming or industrial-chemical agriculture, is disrespectful of nature. In the account given here, humans dominate a natural entity when they exert the preponderance of influence over it. Being a necessary condition for something's existence is not itself to exert such influence over it. Parents are necessary conditions for their children's existence but may not exert such influence over them. Many species on the planet—including those existing in wilderness areas—have human forbearance as a necessary condition for their existence. But this is not to dominate them. Thus that humans are necessary conditions for the existence of some aspects of nature is not necessarily to dominate or show a lack of respect for their autonomy in relation to humanity.

There are other conceptions of respecting the autonomy of nature that, unlike autonomy in relation to humanity, suggest that we can and should respect natural entities and systems because they possess specific properties or capacities. Respect for the autonomy of nature might mean respect for self-organizing, autopoietic systems in nature, or it might mean respect for natural entities and systems that are powerful, active, resistant, or resilient to human-induced changes. A wild river actively and powerfully resists human attempts to change its course or flow. Rainy eastern North America is much more resilient in the face of human impacts than is the dry west, and a granite mesa is more resistant to the mountain biker than the fragile desert that lies around it. If respecting nature's autonomy means respecting natural entities and systems in virtue of these properties, then, implausibly, those dimensions of nature that lack these features would not be worthy of such respect. Calling for respect for nature's autonomy in these senses would lead to other counterintuitive results as well, for it is the less powerful, more delicate, and more easily influenced aspects of nature (those natural entities and processes that are less autonomous) that need greater protection. One virtue of respecting the autonomy of nature in relation to humanity is that it does not discriminate in these ways between natural entities.

Nature Influencing Humans

Whether humans dominate a natural entity or process (and in this sense fail to respect its autonomy in relation to humanity) depends not just on the sheer amount of influence humans have over it but also on whether that entity/process influences us in return. Consider that, although spouses exert a high degree of influence over each other, they typically do not dominate each other. That same amount of influence exerted over an acquaintance would likely be considered domination. We are less likely to judge that a high degree of influence of one over another is domination when there is significant influence in return.[5] The autonomy of nature is thus dependent not just on the amount of influence humans have over nature but also on the amount of influence nature has over humans. When a natural system or entity plays an important role in what happens in human culture, that is, when it exerts significant influence over our lives, then substantial human influence over that natural entity is less likely to count as domination and more likely to be compatible with respect for nature's autonomy in relation to humanity than when the natural entity exerts little or no influence over humanity. In human affairs, it is a sign of a healthy relationship when two people exert significant nondominating influence over each other. Such influence is a similar sign of health when present in the human relationship with nature.

This mutual influence dimension of respect for nature's autonomy can help us see certain types of human relations with rural nature in a positive light. Contrast human interaction with rural landscapes and with wilderness areas. Many preservationists would argue that humans significantly dominate rural lands, while wilderness has autonomy from humanity. This need not be the case, given the above account of autonomy. While it is obviously true that wilderness is less influenced by people than are rural lands, it does not follow from this that wilderness is more autonomous. For rural nature significantly affects people's lives in a way that wilderness does not.[6] Often, people who work with rural nature live by the rain, the soil, the sun, the animals, and the plants. Instead of letting the banker, boss, or stock market determine their lives, they let the seasons, temperature, and presence or absence of predators or pests determine their lives. Rural nature can preserve its autonomy in relationship with humanity even when significantly influenced by humans because it can significantly influence us in return.

It is true that a farmer might have machinery, chemicals, irrigation systems, greenhouses, insurance, and so on, so that she is hardly more influenced by nature than is an urban dweller. But some farmers put their livelihoods in the hands of nature. They depend on the rain coming instead of irrigating with fossil water. They depend on insect predators in the hedgerows instead of chemical pesticides. They depend on the hawks to keep the field mice down. They depend on horses to plough and manure the fields. By leaving themselves open to significant influence from natural entities and systems, their relation with nature is likely to be a nondominating one.

One implication of this account is that if we want to respect the autonomy of nature, it helps not to protect ourselves too much from it.[7] We can sometimes work toward a respect for nature's autonomy by leaving or making ourselves vulnerable to nature.[8] Leaving or restoring predators is one way to accomplish this. When rural people must take down their bird feeders, properly seal their garbage, hike with bells, or give up certain trails because of bears or cougars, they are vulnerable to nature's influences and thus more likely to relate to an autonomous nature. Restoring to the rural landscape wolves that might eat our sheep forces us to change our grazing practices, adds to nature's influence over our lives, and lessens our control of the situation; thus it likely increases the autonomy of local nature in relation to humanity. When humans accommodate themselves to natural processes and entities rather than reworking or eliminating those processes or entities, they show a respect for the autonomy of the nature with which they live.[9]

Virtues of an Ethic of Respect for Nature's Autonomy

Thus, unlike a pure preservationist ethic of noninterference, respect for nature's autonomy in relation to humanity gives us some guidance for how to treat the mixed or rural landscape. The only guidance pure preservationists offer for our treatment of the nature we use is to minimize our involvement with it (or use it as efficiently as possible). As Eric Katz says in his contribution to this volume, "Even in the case of hybrid environments, we ought to lean toward leaving nature alone." But then our advice to the farmer, the forester, and the rural homebuilder is to do as little farming, forestry, or home building as possible. If respect for nature means leaving nature alone, then using nature involves disrespecting it, and at best we can minimize our disrespect by using it as little as possible. In contrast, if respect for nature can mean respecting its autonomy in relationship with humanity, then it is possible to use nature while respecting it. Use of nature that does not compromise its autonomy can be respectful use.

Not only does a pure preservationist ethic give no real guidance for our treatment of less than fully wild nature, but it tends to disparage the value of that type of nature. Lack of respect for less than fully wild nature is an increasingly frequent target for critics of the pure wilderness preservation ethic. Val Plumwood puts the criticism this way: "A dualistic wilderness cult which confines respect and the status of 'nature' to pure virgin land does not support a culture of respect for ordinary land or for nature in the context of everyday life" (Plumwood 1998:667–668). By making respect for the autonomy of nature in relation to humanity a central focus of an environmental ethic, we can avoid viewing nonwilderness lands and humanly influenced species as inferior, degraded versions of wild nature. Rural lands and domesticated animals and plants—though more greatly influenced by humans—can be just as autonomous as wild nature. We can respect them by influencing them in a way that does not dominate them and letting them influence our lives in return.

A purely preservationist ethic tends to define nature par excellence as wild nature or wilderness, entities or places devoid of a high degree of human influence. Once natural entities have been significantly influenced by humans, they lose their status as nature or natural. Thus, for Rolston, a nature improved upon is no longer real nature, and for Katz and Keekok Lee (1999), significantly human-influenced biota are artifacts, whether they be restored landscapes or domesticated animals or plants. But it is not plausible to claim that a formerly barren lake in which humans introduce fish is no longer nature, and neither is it plausible to claim that replanted forests, horses, or cattle are nonnatural, human-created artifacts, as artifactual as plastic chairs. Granted, these entities are not wild nature, but, unless unspoiled wildness is one's criterion for nature, things can be nature and natural, without being wild nature.

The concept of the autonomy of nature in relation to humanity can help us here as well. We can argue that human influence over landscapes or other natural entities need does not render them nonnatural, artifactual beings as long as they retain autonomy in their relation to humanity. In this account, some rural landscapes and some domesticated animals and plants can continue to count as nature and natural, even though they are significantly influenced by humans. Only when their relation to humanity is such that their autonomy has been undermined does it become plausible to argue that they are no longer natural but rather artifactual beings who belong to the category of culture more than to nature. When nature is not defined as the absence of humans or human influence but as requiring the absence of human control and domination and the presence of autonomy in relation to humanity, then less than fully wild nature can be natural and remain real nature

One example of the excesses of defining nature as wild nature is Keekok Lee's claim that "transgenic organisms are artifacts with a degree of artifacticity analogous to that of plastic toys" (1999:53). Inserting a few genes into an organism with tens of thousands of genes hardly turns it into a human-created artifact analogous to a cultural phenomenon like a hula hoop. Similarly, a replanted forest, or even a vegetable garden, retains sufficient autonomy from humanity to qualify as nature. Sun, rain, birds, bugs, and all sorts of natural processes continue to operate beyond human control in gardens and forests, giving them a plausible claim to autonomy from humanity. A fish tank or bonsai garden, on the other hand, may be sufficiently under human control and artifactual that the label "nature" may be more plausibly withheld. By defining nature not as the absence of humans or human influence but as requiring the absence of human control and domination and the presence of autonomy in relation to humanity, we allow for a human place and role in nature. Humans can use nature and natural entities without necessarily destroying their essential character. Our use of nature can be respectful of it, provided it retains its autonomy in the context of this use. The strict separation of humans and nature (i.e., human/nature apartheid) need not be our only way of respecting nature. Human participation in nature and involvement with natural entities, constrained by respect for their autonomy, are equally important components of an ethic of respect for nature.

Conclusion

While preservationist intuitions and policies are of crucial importance to a proper respect for nature, by themselves they provide a merely negative model of humanity's relationship with nature. By supplementing preservationism with an ethic of respecting nature's autonomy in relationship with humanity, we allow for a positive role of humans in nature. With respect for the autonomy of nature as a central moral norm for the human relation with nature, human involvement with nature need not be harmful or degrading to nature in this important respect. This opens the door to the respectful human use of nature and to humans flourishing in nature as real possibilities.

NOTES

1. In addition to the examples in the text, see Robert Elliott (1997), Ned Hettinger and Bill Throop (1999), and Keekok Lee (1999).
2. Taylor explicitly allows that some interference with nature is compatible with respect for nature (1986: 94). But most of his examples involve humans undoing damage they have caused, and this is not an overall positive involvement with nature. He also gives examples of medical assistance to wild animals and providing food and housing for birds. But even if these examples can be made consistent with Taylor's fundamental duty of noninterference, they are not examples of respectful human use of nature.
3. Callicott (1991) is right about the importance of envisioning such a notion, which he calls "sustainable development." Unlike Callicott, I believe that such a relationship should go hand in hand with an ethic of wilderness preservation and need not displace it.
4. For a discussion of these notions, see Rolston (1988:197–201; 1994:181–184) and Lee (1999:177–180).
5. Of course, mutual domination is possible. But a high degree of influence of one party over the other that would be considered domination absent a corresponding influence in return need not be a situation of mutual domination when the corresponding influence is present.
6. It is true that when a person hikes or camps in wilderness, nature has great influence over that individual, and the individual has little or no influence on the wilderness. But such influence is temporary. It is a kind of vacation influence and much less powerful and long term than the influence rural nature has over the lives of rural people who live with and by nature.
7. I thank Bill Throop for this idea specifically and for discussions that helped enormously in the genesis of the ideas in this essay.
8. Examples where people are vulnerable to nature and have little choice in the matter—lightning strikes, tornadoes, the inevitability of death, etc.—provide particularly powerful evidence for the idea that nature is not completely under the human thumb. In these respects, nature might even be seen as dominating humans and thwarting our autonomy. See Katz (1997:133–146) for a discussion.
9. Michael Gill has raised a worrisome counterexample to the suggestion in this section. If making ourselves more vulnerable to nature is a step toward respecting nature's autonomy in relation to humanity, then human-caused global warming—a dramatic human-caused alteration of nature if there ever was one—should be seen as a step in the direction of respecting nature's autonomy, for global warming may well make us more vulnerable to nature, and it is a clear example of nature influencing us in return. Other examples raise the same worry: Are humans respecting nature's autonomy when they clear-cut hillsides and make their homes and villages more vulnerable to massive landslides? Are we respecting nature's autonomy by suppressing fire so that we are more vulnerable

to massive landscape-scale fires? I am not sure what to say about these examples. On the one hand, these examples of human influence over nature that makes us more vulnerable to it illustrate the point that humans are not in control of nature in these situations. That when we dramatically influence nature we often do so at our own peril suggests that we are far from dominating nature (despite trying to). Nature remains autonomous from us even in such cases of significant and harmful human influence. On the other hand, such human activity and nature's response hardly constitute healthy human-nature interaction, and increasing human vulnerability to nature by such dramatic influence is not a way of respecting nature's autonomy in relation to humanity. A comparable human analogue might be driving one's spouse to mental instability with the result that one's life is greatly and negatively affected in return. Perhaps these examples should count as examples of mutual domination and not as a type of mutual influence that lessens the likelihood of domination. These examples suggest that the intentions of humans whose activity increases their vulnerability to nature may play a role in whether one should characterize this activity as a step toward respecting nature's autonomy in relation to humanity. They clearly show that increasing human vulnerability to nature is not a sufficient condition for acting in a way that shows respect for nature's autonomy in relation to humanity.

REFERENCES

Callicott, J. Baird. 1991. "The Wilderness Idea Revisited." *Environmental Professional* 13: 235–247.

Elliot, Robert. 1997. *Faking Nature: The Ethics of Environmental Restoration.* London: Routledge.

Hannah, Lee, David Lohse, Charles Hutchinson, John L. Carr, and Ali Lankerani. 1993. "A Preliminary Inventory of Human Disturbances of World Ecosystems." *Ambio* 23: 246–250.

Hettinger, Ned and Bill Throop. 1999. "Refocusing Ecocentrism: De-emphasizing Stability and Defending Wildness." *Environmental Ethics* 21: 3–21.

Kane, Stanley. 1994. "Restoration or Preservation? Reflections on a Clash of Environmental Philosophies." In A. Dwight Baldwin, Judith de Luce, and Carl Pletsch, eds., *Beyond Preservation: Restoring and Inventing Landscapes*, pp. 69–84. Minneapolis: University of Minnesota Press.

Katz, Eric. 1997. *Nature as Subject.* Lanham, Md.: Rowman and Littlefield.

——. 2005. "The Liberation of Humanity and Nature." In Thomas Heyd, ed., *Recognizing the Autonomy of Nature: Theory and Practice*, pp. 75–85. New York: Columbia University Press.

Lee, Keekok. 1999. *The Natural and the Artefactual.* Lanham, Md.: Lexington.

Plumwood, Val. 1998. "Wilderness Skepticism and Wilderness Dualism." In J. Baird Callicott and Michael P. Nelson, eds., *The Great New Wilderness Debate*, pp. 652–690. Athens: University of Georgia Press.

Rolston, Holmes III. 1988. *Environmental Ethics*. Philadelphia: Temple University Press.

——. 1991. "The Wilderness Idea Reaffirmed." *Environmental Professional* 13: 370–377.

——. 1994. *Conserving Natural Value*. New York: Columbia University Press.

Taylor, Paul. 1986. *Respect for Nature*. Princeton: Princeton University Press.

Turner, Jack. 1996. *The Abstract Wild*. Tucson: University of Arizona Press.

Visvader, John. 1996. "Natura Naturans." *Human Ecology Review* 3 (autumn): 16–18.

SIX

Autonomy and Agriculture

WILLIAM THROOP AND BETH VICKERS

Leopold's Distinction

A LITTLE over fifty years ago, Aldo Leopold distinguished between two approaches to land use, which he characterized as "the A/B cleavage." Type A agriculture and forestry attempted to maximize yield using the best science available, while type B aimed at using the land in ways that preserved the health of the whole ecosystem. In the former, humans saw themselves as ruling over the land, whereas in the latter humans were conceived as "plain members and citizens of the biotic community."[1] Although Leopold condemned the type A approach on both practical and moral grounds, he did not develop a moral critique of the domination that essentially characterizes it. Others have argued that the domination of nature is in itself morally problematic,[2] but such critiques have not shown that we can draw a Leopoldian distinction between domination and respectful participation in an ecosystem. They do not show how type B agriculture and forestry can avoid dominating nature while still radically transforming ecosystems. An analysis of domination will not provide a positive image of how we should pursue agriculture and forestry; for this, we must have an account of the value preserved when domination is absent. We think that the term "autonomy" captures key dimensions of this value. In humans, autonomy is the value that is compromised through domination. We suggest that it seems natural to extend this to nonhumans when we seek to understand why dominating them is wrong.

Of course, Leopold did develop an ethic for land use that provides the kind of positive image we seek. Agriculture that preserves the integrity, stability, and beauty of the biotic community would be appropriate or "right," and that which compromises these features of ecosystems would be wrong. This ethic has encountered serious challenges, however. Increasingly, Leopold's focus on stable ecosystems has seemed misguided as ecological theory has shifted to a paradigm that emphasizes the dynamic nature of ecosystems.[3] Moreover, neither Leopold nor his followers have explained how we should understand the integrity of ecosystems that have been heavily modified by humans, such as agricultural systems. The account of ecosystem autonomy that we develop below can be seen as a reinterpretation of Leopold's ethical theory that makes it compatible with contemporary ecology and highlights the moral value associated with working with nature rather than dominating it.

We begin by briefly indicating why we reject Eric Katz's account of ecosystem autonomy. We then develop an alternative account of autonomy and show why it should carry significant moral value. Although violating autonomy, in our sense, is prima facie morally wrong, we do not take our analysis to show that land use that dominates ecosystems is always wrong. In some cases, humans must violate the autonomy of nature and its inhabitants. However, the burden of proof lies heavily on those who would violate autonomy significantly when alternatives are easily available. We explore some implications for agriculture of ecosystem autonomy; similar points could be made about forestry. We consider several objections to our account and suggest ways in which it must be further developed. Although we cannot develop a full explanation of ecosystem autonomy in this paper, our account aims to move the discussion beyond the merely metaphoric extension of the concept.

Autonomy as Wildness

Eric Katz has developed the most influential account of autonomy for ecosystems in the environmental literature. A central theme in his *Nature as Subject* is that "nature is an autonomous moral subject, analogous to the traditional human subject in moral theory."[4] He maintains that natural systems are autonomous when "they are independent from external design, purpose or control."[5] As a result, for Katz, when ecosystems are characterized by human purposes, their autonomy is compromised. They are artifacts, and, as such, they manifest human domination. Since such domination is morally problematic, we should avoid it, where possible. Since wildness is the property of ecosystems when they are unfolding on their own, respect for the autonomy of ecosystems should lead us to promote wildness wherever we can.

One can extend the notion of autonomy to ecosystems in a variety of ways. Each way has advantages and disadvantages, but we should not expect a unique correct way to emerge. Such extensions should be evaluated on the strength of their connection to the original concept and on their utility. Katz's account of autonomy has the advantage of giving us clear guidance in many cases, and we believe that it does make "autonomy" an important value-bearing term. It also has two unfortunate implications, however, which lead us to think that an alternative account is preferable.

First, Katz's view leaves unfulfilled the promise that an account of autonomy can tell us how to develop a positive relationship with nature. If ecosystem autonomy is defined in terms of total independence from human control, then respect for autonomy requires us to leave nature alone whenever possible. This tells us what not to do, but it does not tell us how to relate to land we must use (except to use it as little as possible). Our only guide for relating to rural landscapes, which by their nature reflect human purposes, is to back off. This limits the usefulness of appeals to autonomy, for it does not help us to distinguish between the moral value of different agricultural practices.[6] Leopold's distinction suggests that we can work with nature in a way that does not compromise it, yet Katz leaves little room for humans to adopt the role of "plain members" of the biotic community.

Second, his account divorces ecosystem autonomy from human autonomy and thus creates a much thinner connection between a land ethic and standard ethical traditions in the West. A human can be autonomous even though his/her character has been significantly influenced by others, yet, for Katz, ecosystem autonomy is violated by analogous influence. For example, I do not compromise a friend's autonomy by trying to get him to serve one of my purposes, say, by giving me a ride somewhere. And I do not compromise a student's autonomy by shaping her writing to reflect proper academic standards. Of course, the friend consents to the ride, and perhaps the student has consented to be shaped by enrolling in the course. It is far from clear, however, that consent is required for influence to be compatible with autonomy in humans; when one teaches one's children, one often does not have their consent. More important, even if some kind of consent is usually present when one's influence over another human is not domination, it does not follow that consent would be necessary for influence to be compatible with respect for ecosystem autonomy. As we will see, consent may just be peculiar to the form of self-governance appropriate to rational beings. We should expect that the extension of autonomy to ecosystems will require significant reinterpretation of the concept, but presumably a reinterpretation that keeps its basic structure.

Intuitively, I need not compromise an ecosystem by shaping it to satisfy some of my purposes. It seems implausible to say that a northern hardwood forest ecosystem's autonomy is compromised by careful selective logging. If the logging mirrors common disturbance regimes, the system remains governed by the same principles that characterized it prior to the logging. Intuitively, ecosystems that incorporate populations of humans living sustainably seem to have their natures characterized in part by the contributions humans make to the natural cycles. We suggest that autonomy can guide an agricultural ethic only if it can interpret some human influence as being in accord with the nature of an ecosystem.[7]

The Autonomy of Ecosystems

Why emphasize ecosystem autonomy in an agricultural ethic? Such an ethic seems promising for two reasons. First, in a Kantian ethic, respect for the autonomy of human beings is a central value. An agricultural ethic based on the autonomy of land would extend powerful ethical norms beyond their normal human sphere. If sense can be made of ways in which an ecosystem can be autonomous, then we would have reasons to respect the autonomy of ecosystems and limit our actions accordingly. This appeal to autonomy would explain why many environmentalists are concerned about domination of the land. Second, since humans can be autonomous even when they are subject to significant external influences, the metaphorical extension of autonomy to ecosystems should permit some kinds of human influences and prohibit others. By developing an account of the differences between human influences that compromise the autonomy of some agricultural land and those that do not, we provide one part of a positive agricultural ethic.

A natural objection to the extension of autonomy beyond humans is that autonomy typically involves free choice. It is a characteristic possessed by some humans by virtue of their ability to prescribe laws to themselves. Since ecosystems are nonrational entities, they cannot be autonomous in this sense. But if we take the etymology of autonomy seriously; it is self-rule: its opposite is being ruled by another. Generically, to be self-ruled is to have one's behavior regulated by one's own principles. For most mature humans, "one's own principles" are those that one chooses. For beings that do not have free choice, "one's own principles" must be construed in terms of the principles that characterize the nature of such beings.

Let us stipulate, then, that one's actions are compatible with respect for the autonomy of an entity that cannot exhibit free choice if and only if they are compatible with that entity's behaving in accord with its nature. This concept of autonomy can be applied to a variety of kinds of entities; for example, a tree could exhibit autonomy.[8] Allowing an apple tree in the garden to manifest its species-specific form respects its autonomy, whereas growing it as an espalier does not. People can recognize trees from a distance because they know the forms that trees normally take when conditions for their growth are adequate. Similar things apply to most species, though one may need a great deal of expertise to understand the nature of many species, and the presence of borderline cases will make it impossible to identify the nature of some entities.

Even if we can make sense of extending autonomy to nonhuman organisms, we encounter more significant challenges when we extend it to ecosystems, for ecosystems do not seem to be differentiated into natural kinds. Over many spatial and temporal scales, nature seems best characterized by change. With the rise of what Donald Worster calls "the ecology of chaos," the image of ecosystems moving along standard trajectories toward climax states has been modified beyond recognition.[9] As a result, we cannot identify the way an ecosystem is supposed to be in a region. The mind's independent world does not fix the identity of the ecosystem. Consequently, it would appear problematic to talk about the nature of an ecosystem in a region as if it were a discrete biological entity. If we are to extend autonomy to ecosystems, we must give an account of the nature of an ecosystem in a region that accords with current trends in ecology. We do speak colloquially about specific kinds of ecosystems, for example, a northern hardwood ecosystem or an oak savanna, but are we simply imposing artificial categories on chance assemblages of species, or are we justifiably identifying systems that, while fluid and hierarchically ordered, have enough reality to be characterized by autonomy? We believe that by applying a pragmatist account of truth to statements about the nature of ecosystems, we can provide the outlines of a plausible account of ecosystem autonomy.

Over the last two decades, Hilary Putnam, among others, has been defending a version of pragmatism that understands truth in terms of ideal warranted assertability that sees facts and values as inextricably intertwined and recognizes a plurality of legitimate ways of characterizing the way the world is.[10] In this view, the nature of an entity is a function of both the empirical data we get from the world and our interests. As Putnam cryptically puts the point, "mind and world jointly make up mind and world."[11] Those who accept a correspondence theory of truth may believe that there is a uniquely correct view that characterizes the world as it is, independent of perceivers. In such a view, the nature of an apple tree is determined by the world itself, and so presumably would be its autonomy. According to a pragmatist theory of truth, "apple trees have alternate leaves" is true if we would be warranted in asserting it in ideal circumstances for assessing such claims. The norms for warranted assertability and for ideal circumstances are provided in part by us, and, since they depend on our interests, different claims may be acceptable for different groups. We leave the defense of this view of truth to Putnam and others,[12] and we take no position on whether it applies to all sentences. It seems particularly useful for sentences that include terms such as "ecosystem" since such terms do not appear to "cut the world at its joints." In brief, our view is that the nature of an ecosystem in a region depends in part on the current ecosystem processes and in part on our interests, which determine our norms for ecosystem degradation. We cannot simply decide what ecosystems exist in an area by consulting an interest-free science; our interest-laden norms play a role in determining the nature of an ecosystem and hence its autonomous functioning. Of course, pragmatists think this is true for all categories, apple trees as well as ecosystems, but by focusing on our role in determining the nature of ecosystems, we can see why a thoroughly dynamic view of nature need not undermine an account of ecosystem autonomy rooted in an ecosystem's behaving according to its own nature.

We now turn to the details of one promising way in which a pragmatist could specify the nature of an ecosystem. In this account, the nature of an ecosystem is defined in terms of the current state of the system and an idealized extension of our knowledge of ecosystem functioning and our norms for evaluating ecosystem degradation. *If, given our (idealized) knowledge of ecosystem functioning and our (idealized) norms for ecosystem degradation, we are warranted in asserting that an ecosystem is above a threshold for degradation, then, other things being equal, the current state of the ecosystem defines its nature. If we are warranted in asserting that the system is below the threshold, its nature is defined in terms of either a prior state of the system that was above the threshold or a suprathreshold state that would evolve from the current state under conditions of reduced human impact.*

We do not know what we would believe about ecosystem degradation in ideal circumstances; our understanding of ecosystem dynamics is in its infancy. It seems unlikely, however, that the following crude norms will fail to be justified: other things being equal, a system is degraded if it loses complexity, if it loses resilience, if it loses a kind of stability that it had, if it loses diversity, and so on. Minor losses may constitute degradation but may leave us with a system that still functions reasonably well. Along a continuum of degrees of degradation, we can mark a region beyond which degradation causes dramatic ecosystem alterations; call this a threshold for degradation. Where an ecosystem is below the degradation threshold, we have a choice about how to identify an ecosystem. The degradation threshold for an ecosystem depends on our norms and knowledge, as well as on the status of other systems. For example, a lower threshold might be reasonable if significant degradation is widespread. In this respect, assignment of a threshold for degradation mirrors the assignment of minimal conditions for autonomy in humans. What we value as autonomy in humans is in part a function of our beliefs about human capacities for self-rule. In a world where this capacity is quite limited, say, a world of preteenagers, the threshold for autonomy would be lower than in our world.

We take this account of the nature of an ecosystem to reflect ordinary practices. We tend to identify an ecosystem with its current state unless we have some reason not to do so.[13] We acknowledge a degree of arbitrariness in this identification, but the unease that may accompany this acknowledgment is reduced by our recognition that many kinds of identity are a function of human interests, especially the identities of collective entities such as families and communities. We cannot identify a system in any way we want; we are constrained by the system itself and also by our intellectual and social traditions. The latter may be criticized, but only from the vantage point of some other acceptable norms or information. Thus the arbitrariness is limited. Our specification of the current state of the ecosystem and the threshold for degradation reflects the interplay between our interests and the empirical facts. The current state, or some prior predegradation state, will often be specified in quite general terms, unless we have particular reasons for being more specific.

For example, we may identify a system as a mixed community of small farms and northern hardwood forests, rather than in terms of specific species composition. When we talk of a current state, we do not mean a snapshot of the flora and fauna; rather, we designate the characteristic ecosystem processes in an area. The threshold for degradation should also be interpreted as a general region on the continuum of degradation rather than as a definite line. Radical shifts in species composition and significant loss of resilience are indicators of an ecosystem crossing a threshold, but we should not anticipate precise accounts of how much shift or loss is necessary for the threshold to be considered crossed. Generally, thresholds are marked by an ecosystem moving on a trajectory distinct from the trajectories that characterized the prior system. For example, when a lake has a large enough increase in nutrient loading, it may shift from an oligotrophic system to a eutrophic system. Although some thresholds are fairly clear, others will be subject to debate. With the latter, we would expect that evaluations of when the threshold has been passed should be subject to rational adjudication in the long run, as we acquire more understanding about ecosystem functioning. Naturally, in the short run, political factors may play a significant role in settling such disagreements. In the absence of evidence that is sufficient to establish where a threshold exists, we should exercise caution when approaching possible thresholds, avoiding the potential for violating autonomy where possible.

The above account of the nature of an ecosystem in a region needs significant refinement, but it should be sufficient to show how we can legitimately segment the flux of nature into identifiable ecosystems. This done, we can say that human activity that accords with the nature of an ecosystem in a region (i.e., activity that does not alter that nature) respects that ecosystem's autonomy. Human activity that significantly alters an ecosystem's nature violates its autonomy. One consequence of this account is that reasonable people may disagree about exactly when significant ecosystem autonomy has been lost. We must rely on our current knowledge to make such assessments, but we must be ready to revise our appraisal in light of new evidence. Our assessment of autonomy will be multicriterial, like our assessment of human autonomy. In addition to the multiple norms for degradation, heuristic guides to loss of autonomy will inform judgments. Do humans operate as "plain members" of the ecosystem community, as Aldo Leopold urged? Does the community have a history that reveals a resilient ecosystem? Has human management taken into account the perspectives of many other members of the community? Have we listened to the land and been influenced by it? Positive answers to such questions suggest that we are respecting ecosystem autonomy, because such factors should play a role in the justification for norms for degradation.[14] The uncertainty that characterizes controversial cases should not blind us to the many clear cases where we have violated ecosystem autonomy. In the latter cases, the appeal to autonomy provides a powerful argument against land use practices, including, we believe, many of those used in corporate agriculture.

Another consequence of this account is that ecosystem autonomy can be compromised by nonhuman events and hence without human domination.[15] An ecosystem may be significantly altered by a volcanic eruption and thereby have its autonomy compromised. We may not wish to say such natural events are bad, all things considered, and they do not seem to involve domination, but they are certainly bad for the ecosystem: they harm a great many ecosystem members and significantly alter ecosystem processes.[16] Similarly, human autonomy can be compromised by nonintentional events, such as Alzheimer's disease, without any implication that a human is being dominated. In this case, as with the volcanic eruption, the entity is no longer self-governing. If, however, we compromise the autonomy of a system, then we are dominating it. Only the latter has moral implications, because only it is caused by moral agents.

Valuing Ecosystem Autonomy

Should autonomy so construed be given moral value? We believe so; our obligation to respect ecosystem autonomy coheres well with a variety of widely held empirical and normative beliefs. The coherence theory of justification for moral values has been well defended elsewhere, and we will not outline its merits.[17] One of the problems with providing a coherentist justification for a specific moral belief is that it is impossible to outline fully the range of considerations with which the belief coheres. We mention just a few beliefs that fit well with the value of autonomy in ecosystems.

First, the extension of autonomy to ecosystems can be viewed as the end result of a series of natural extensions. We can talk of autonomy in the developmentally disabled even though their powers of rational choice may be quite limited. Many caregivers believe that the developmentally disabled should have control over their lives except where they threaten injury to themselves or others. If we countenance an application of autonomy to humans with very limited capacity for rational choice, then the extension of autonomy to animals is quite natural. Indeed, this extension explains fairly common intuitions. Many see something degrading in the training of dancing bears for a circus. The bears are being made to do things far from their natural tendencies. Objections to zoos are sometimes based not just on physical harm to animals but on loss of autonomy.[18] We have already mentioned the extension of autonomy to plants; intuitively, espaliers and topiary do not accord with the nature of trees so shaped. Here, we presume that the pain of not being able to act in accord with one's nature is not at issue; rather, it is the violation of that nature. But how bad is the violation of the autonomy of a nonconscious entity? Suppose one marks the bark of a tree or tattoos a friend's arm; these are apparently behaviors that alter the nature of an entity. The extent of the alteration and the moral considerability of the entity would bear on the disvalue. The wrong may be less severe in the case of nonconscious beings like apple trees, but if they have some moral considerability, then it is nonnegligible.[19]

So far, these extensions are to individuals, but it is also natural to talk about violating the autonomy of groups, say, nations or corporations.[20] One nation can oppress another by preventing its self-rule in key areas. Many people believe that international organizations such as the United Nations can unacceptably violate a country's national autonomy. While we may disagree, we find the discussion of national autonomy quite natural. Once we extend autonomy to nonrational individuals and to groups, it seems a very short step to extend it to ecosystem communities; indeed, it seems somewhat arbitrary to deny such extension. The above range of cases strongly suggests that self-rule is the basic value and that the familiar concept of human autonomy—rule by one's own (uncoerced) choices—is the manifestation of that value that is appropriate in rational beings. If so, the moral force of human autonomy is in part attributable to the basic value.

Second, evolutionary and ecological theories imply that humans have evolved like other organisms and share with them membership in ecosystem communities characterized by cooperation and competition. Although these empirical assumptions do not have immediate normative implications, Paul Taylor argues persuasively that they cohere with value systems that extend moral value beyond the human sphere.[21] Once we accept a worldview in which other animals are teleological centers of life with whom we bear complex evolutionary and ecological relations and we recognize that the arguments for believing humans have a special place in the world seem based on shaky foundations, we should find these views in serious tension with any normative theory that gives only humans moral considerability. Coherentist extensions of some moral value need not be accompanied by the extension of the value of autonomy, but the arguments run in parallel, and it seems arbitrary to extend some values and not others where such parallelism exists. We believe that a coherent theory that includes contemporary evolutionary theory and ecology would extend attributions of moral value related to harm and to autonomy to nonhumans. It is a further move to extend these values to systems of individuals, but if, as we have suggested, we can make sense of extending autonomy to such systems, then it would be arbitrary to do so without also extending the value associated with autonomy. Indeed, it is the value of autonomy that seems to drive the extension of autonomy to human groups such as nations.

Third, arguments based on the differences between humans and ecosystems are unpersuasive and reveal unjustifiable biases against nonhumans. Suppose, for example, one argues that humans are harmed by loss of autonomy whereas ecosystems are not, so ecosystem autonomy is not a moral value. But what is harm? Broadly construed, harm is a function of our norms for how an entity ought to exist. If we recognize norms regarding ecosystem degradation, then we have norms regarding how an ecosystem ought to be (in a given context). This permits us to extend the notion of harm to ecosystems; an ecosystem is harmed in the relevant sense if it is degraded. That the norms are partially a function of human purposes and historical circumstances is no argument against the pragmatist's assessment of ecosystem harm. The same can be said of human harm, although the latter may be more robustly rooted in a broadly accepted view of human nature. One can, of course, reserve a term for a narrow construal of a concept. One can restrict the word "harm" to sentient beings and "autonomy" to rational beings, for example. But then one can invent cousin terms, which carries similar moral weight, to cover the broader category: "shmarm" and "shmautonomy," perhaps. What seems arbitrary in these cases is the restriction of the value to the narrow term.

Respecting the Autonomy of Agricultural Land

What are the implications of valuing the autonomy of agricultural land? We cannot move directly from a general value to specific acontextual obligations. We can only sketch general directions in which we might be obligated to move (or kinds of policies we might adopt) and recognize that specific obligations will depend on contextual factors. Given a few plausible assumptions, our analysis of autonomy supports the following two general claims. First, a strong burden of proof lies on the defenders of industrial agriculture because its practices frequently violate the autonomy of the ecosystems in which they are exercised. Second, where land has a significant and generally benign agricultural history, we should support small-scale, community-focused agriculture, economically and socially, and we should attempt to make it economically feasible for more small family farmers to practice community-focused agriculture.

The first claim depends on an argument that industrial agriculture frequently leaves the environment below the degradation threshold. Industrial agriculture typically intensively uses fertilizers and pesticides, and it monocrops in ways that severely limit the other organisms that can flourish in less intensively farmed ecosystems. Its increasing reliance on genetically modified organisms poses additional threats to biodiversity. According to our norms for ecosystem degradation, these results often count as degradation. In addition, industrial agriculture negatively affects the human community structure that characterizes most small-farm agricultural landscapes. Since one degrades a mixed ecosystem of humans and other organisms if one eliminates the human communities that allow it to thrive, such community alterations count as ecosystem degradation. In light of the severity of these forms of degradation, we believe that most industrial agriculture often falls below the degradation threshold, and hence it does not define the nature of an ecosystem. If we are right, then typically industrial agriculture constitutes morally problematic domination of the land; it violates the autonomy of the land. Other values may outweigh the value of ecosystem autonomy and thus justify industrial agriculture. The argument that it is needed to feed the growing human population is one such case. We doubt the validity of this argument, but space does not permit us to develop our critique of that line of reasoning here. The above considerations suggest that industrial agriculture faces a significant moral hurdle. If industrial agriculture does significantly degrade ecosystems, then it does violate their autonomy (given our analysis of ecosystem autonomy). Thus its defenders must shoulder the burden of showing that a significant moral wrong is justified by extraordinary circumstances; it is not enough to argue that industrial agriculture is just a more efficient and cheaper way of producing food than the alternatives. Such considerations typically do not trump moral values.

The second claim depends on both a characterization of community-focused agriculture (CFA) and on some information about regional histories. CFA typically exists on relatively small scales (from a few acres to a couple of hundred) and supports and engages a local human community, often by providing fresh meat and produce for retail sales and supplying local restaurants. In some versions, communities have created cooperatives called CSAs (community-supported agriculture) in which a seasonal fee is paid to the caretakers of a farm in exchange for certain amounts of fresh produce grown and picked throughout the season. Compared to other contemporary agricultural systems in the developed world, the cycles that characterize CFA are relatively closed. A large portion of the produce stays within the system, as does the waste produced. The human inputs from outside the system are minimized. Many contributors to the small-scale organic agriculture movement provide modern exemplars of CFA.[22]

A look at the history of land now being industrially farmed indicates a significant portion of it has been characterized by CFA systems. This is especially true of our region, in the northeastern United States, but it is also true of many other areas in the world. While not all of these systems were above the degradation threshold, many were, and these partially characterized the identity of the ecosystems in which they figured. In other words, we believe that according to our (idealized) norms for ecosystem degradation, many early landscapes of mixed agriculture and woodlots were not significantly degraded. Thus, according to our account of the nature of an ecosystem in a region, these systems then characterized the identity of the ecosystem there. Of course, the development of these agricultural ecosystems involved the destruction of earlier nonagricultural systems. This leads to temporal-scale questions that we address below. In the last half century, some of this early agricultural land has been transformed again by industrial agriculture, but since this leaves us with a degraded landscape, it does not change the nature of the ecosystem. Often, such industrial agricultural systems can frequently be moved above the degradation threshold by setting them on a trajectory toward a new CFA system. This is not the only healthy trajectory available in most cases; such systems might revert to forest, for example. We need food, however, and we should seek to produce it in a way that respects the autonomy of the land. We believe that typically CFA best supports the autonomy of agricultural land where it has the appropriate history, because it is least likely to degrade the land. Naturally, any agricultural system will degrade land when it tries to provide food for too many people; CFA is no exception. However, CFA typically uses the land less intensively because it provides food predominantly for a local population. It uses local inputs, so it avoids mass-marketed herbicides and pesticides. CFA is also the system that most thoroughly integrates humans into an agricultural ecosystem. People experience the food webs in which they are embedded, and they are influenced by nature as they manage the land. The human and natural communities are deeply interwoven, and as a result residents tend to be aware of ecosystem degradation.

Where small-scale agriculture has a strong historical presence, it characterizes the identity of the landscape (providing it has not systematically degraded the ecosystem). We respect the autonomy of that landscape by supporting an agricultural system that is most likely to leave the system above the degradation threshold. We believe that CFA best represents this promise. Other agricultural systems have characterized other areas without leaving the ecosystems below the degradation threshold. In those areas, respect for the autonomy of the land will have very different implications.

In sum, if rural landscapes are above the threshold, then we should recognize that their autonomous functioning has moral value. We should act in ways that facilitate their functioning in accord with their current nature, unless we have good reason to do otherwise. The burden of proof lies on those who wish to alter an autonomously functioning system. In some cases, burden may be relatively easy to shoulder, especially if the land is being transformed to some other autonomous landscape, perhaps a more rare one. For example, in mid-nineteenth-century Vermont, when most of the state was farmland, returning a farm to forest was probably good, even though it did not maintain the autonomy of the working farmland, for it provided important habitat for indigenous species. Now, however, preserving the remaining small farms and encouraging a community-focused agriculture in the rural northeastern United States seems preferable to further reforestation, since Vermont is now approximately eighty percent forested and preserving the remaining agriculture respects the autonomy of the Vermont landscape. To permit the rise of industrial agriculture to eclipse the traditional farming systems and the communities they fostered would be to violate the autonomy of this landscape seriously.

Objections and Responses

Our defense of the claim that respect for the autonomy of agricultural ecosystems indicts much industrial agriculture and promotes CFA raises a number of issues. Here, we address four natural objections and indicate the directions in which we would push our analysis. First, one might argue that all farming involves loss of autonomy. The change from preagricultural to agricultural society violated the autonomy of the land according to our account, because it altered the identity of the ecosystem. If something once counted as violating autonomy, then it should always do so. A race that has been enslaved for many generations does not acquire an identity as a slave race, and we continue to say their autonomy is violated. Why should ecosystems be different?

We do not believe that history alters the fundamental character of human beings, but it does alter the character of the ecosystems. Ecosystems are dynamic in deeper ways than humans are, or so we believe. Because most humans who are not free retain the capacity for rule by their own choices, we say that autonomy of enslaved humans is violated, not that they become a different kind of being (though of course an enslaved person might lose the capacity for self-rule). The capacity to be a forest does not give an ecosystem an identity as a forest, because an ecosystem has many different capacities that are exemplified in different climatic and use conditions. Because of the fluidity of ecosystems, their identity conditions are fixed to a larger extent by their recent history. We also have a moral image of how we should relate to humans that coheres with our views about human identity. We do not think it right to constrain human choices unduly because human identity is linked to choice. With respect to land, the situation is different. In most ecosystems, humans are justified in altering some land so that they live reasonably good lives on it. This justification for alteration reinforces an account of ecosystem identity that makes identity compatible with some alteration. Here, justification runs from moral theory to identity, rather than the other way around, but this is what one should expect in a coherence theory of justification where what counts is the mutual reinforcing relations between our best morality and our views about identity.

A second objection arises from problems of scale. By shifting scale, one shifts the ecosystem whose autonomy counts in an area; this threatens to make almost any land use compatible with autonomy at some scale. Imagine a farmer reclaiming an old pasture by bulldozing early successional vegetation and reseeding. Although this clearly violates the autonomy of the young woodland that had been growing there, the farmer claims that the autonomy of the larger ecosystem has been preserved. She has preserved the agricultural land that historically characterized the area, and she has not degraded the larger-scale patchwork farm and forest landscape. This claim seems reasonable, but an analogue with industrial forestry does not. International Paper might defend one-hundred-acre clear-cuts by claiming that the land they log is traditionally forest-product producing and that on sufficiently large spatial and temporal scales a few hundred hundred-acre clear-cuts do not degrade the ecosystem. In time, a forest will grow back. The appeal to ecosystem autonomy is vacuous unless we have a nonarbitrary way of identifying the appropriate scales for considering ecosystem autonomy.

The problem of scale bedevils any ecocentric ethic. Ecocentrists should adopt a hierarchical concept of ecosystems according to which ecosystems are nested within larger ecosystems and our purposes play a role in identifying the relevant ecosystem boundaries in a discussion.[23] People with different values and purposes will tend to select different scales, and the disagreements that result will often prove intractable. We can, however, clearly identify inappropriate scales. Someone who defends the current biodiversity holocaust on the grounds that, over geological time, the losses are insignificant is working with an inappropriate scale. We do not (and should not) assess environmental degradation over geological time.[24] We take a pragmatist approach to issues of scale that parallels our approach to norms for environmental degradation. We must begin with our current best norms, subject them to criticism, and be prepared to revise them in the light of new evidence. Since scale norms are highly contextual and purpose relative, we can expect that a wider range of scale will survive criticism and the emphasis on certain scales in a situation will be in part a function of the preferences of stakeholders. Still, rational considerations will be able to rule out certain scales as inappropriate for assessing degradation. We believe that rational evaluation of the scales for assessing degradation of intermediate landscapes will lead us to adopt scales that permit pasture clearing (where sufficient woodland remains) and reject most industrial clear-cutting. A defense of industrial clear-cutting would require use of a temporal scale beyond our typical planning horizon and spatial scales beyond those we can experientially appreciate. These criteria are merely suggestive, however, and much work remains to be done on why we should prefer scales that favor smaller farms and less intensive forestry.

Third, much of our analysis has depended on norms for ecosystem degradation and, implicitly, on contrasting norms for ecosystem health. Given this dependence, some may wonder why we should take a detour into the murky realm of ecosystem autonomy. Perhaps we should work on clearer notions of health and degradation and forget the extension of autonomy to ecosystems. Although we define the less familiar concept of ecosystem autonomy in terms of more familiar notions of degradation, we do not think that the value of the former is more basic than the latter. We take these values to be correlated and the descriptive terms on which the values supervene to be interdefinable. The direction of our analysis depends on our purposes, not on the structure of value. We believe that autonomy carries with it a stronger moral dimension than health and that it explains our intuitive concerns about domination of the land. It is no less murky than degradation, just less common. We see this project, in part, as an exploration of the interconnections between key value-laden terms that can be applied fruitfully to ecosystems.

Fourth, many will take our defense of CFA to be a nostalgic romanticizing of a past that was not as pretty as we suggest and one that is unrecoverable in our current global economy. We cannot go back to earlier community structures in our information age, and we will not stand for the time and energy such a life demands. We agree that much prior CFA did have negative impacts on the land and that the community life it fostered had a dark side associated with poverty, cultural isolation, and disempowerment. We do not want to return to some mythical past; rather, we urge the building of a new CFA that utilizes our best current knowledge and technology and appreciates the diverse communities fostered by modern mobility. We find that many young people are attracted to life in small agricultural communities as a counterbalance to the fast-paced land-impoverished lives they have experienced in urban and suburban settings. They do not want to abandon all the amenities of modern civilization, but they want these embedded in lives containing deeper relations to nature and to human community. A new community-focused agriculture promises such lives and also promises a means of life support that is compatible with the autonomy of agricultural ecosystems.

Conclusion

One of the major challenges facing any account of ecosystem autonomy that parallels human autonomy is to distinguish between influences that are compatible with autonomy and those that constitute domination. Since agriculture is one of the primary ways we influence nature, it provides a natural test for an account of autonomy. We have attempted to show how our account of autonomy allows us to make the distinction in intuitively plausible places, in places that reflect Leopold's A/B cleavage. Few environmentalists doubt that agriculture has been moving in the wrong direction as a result of severe economic pressures and subsidies given to agribusiness. Industrial agriculture can be criticized on a variety of grounds, but few carry the moral weight that we believe to be associated with the violation of ecosystem autonomy. Agriculture also poses a significant challenge to environmentalists, who must develop a positive image of how we can live in nature while sustaining ourselves. We believe that the value of autonomy as applied to rural ecosystems can help to provide such a positive image. We have argued that community-focused agricultural systems do constitute part of the identity of ecosystems in some areas, and, where they do so, respect for autonomy should lead us to support such systems.[25]

NOTES

1. Aldo Leopold, "The Land Ethic," in *A Sand County Almanac* (New York: Oxford University Press, 1949), p. 240.
2. See, for example, Eric Katz, *Nature as Subject: Human Obligation and Natural Community* (Lanham, Md.: Rowman and Littlefield, 1997); and Karen War-

ren, "The Power and Promise of Ecological Feminism," *Environmental Ethics* 12 (summer 1990): 125–46.

3. See Ned Hettinger and William Throop, "Refocusing Ecocentrism: De-emphasizing Stability and Defending Wildness," *Environmental Ethics* 21 (spring 1999): 3–21. The positive suggestions for refocusing ecocentrism on wildness that are defended in Hettinger and Throop (1999) have the same limitations we attribute to Katz's account of autonomy.
4. Katz, *Nature as Subject*, p. xxiv. See also Eric Katz, "The Liberation of Humanity and Nature," in this volume.
5. Katz, *Nature as Subject*, p. 129.
6. We cannot simply differentially evaluate agricultural practices on their degree of invasiveness, for almost all modern agriculture very significantly modifies the preagricultural landscape and hence constitutes serious violation of ecosystem autonomy in Katz's account.
7. We cannot here discuss the advantages and disadvantages of other accounts of ecosystem autonomy, but we believe that versions of the above concerns apply to Ned Hettinger's account in "Respecting Nature's Autonomy in Relationship with Humanity," in this volume.
8. In humans, we often draw the distinction between having the capacity for self-rule (being an autonomous agent) and having one's actions governed by one's principles (acting autonomously). When applying autonomy to nonrational beings, we will use the latter sense of "autonomy." In this sense, when a nonrational being's autonomy is violated, it loses that autonomy (though, as we will see, it may become something else that has a different kind of autonomy).
9. Donald Worster "The Ecology of Order and Chaos," *Environmental History Review* 14 (spring/summer 1990): 1–18; and chapter17 of his *Nature's Economy: A History of Ecological Ideas*, 2d ed. (Cambridge: Cambridge University Press, 1994). See also Daniel Botkin, *Discordant Harmonies* (New York: Oxford University Press, 1990).
10. Hilary Putnam, *Reason, Truth, and History* (New York: Cambridge University Press, 1981); idem, *The Many Faces of Realism* (LaSalle, Ill.: Open Court, 1987).
11. Putnam, *Reason, Truth, and History*, p. xi.
12. For a detailed discussion of pragmatic accounts of truth, see Mark Johnston, "Objectivity Refigured: Pragmatism Without Verificationism," in John Haldane and Crispin Wright, eds., *Reality, Representation, and Projection* (New York: Oxford University Press, 1993). Johnston uncovers several problems of detail with warranted assertibility accounts of truth and thereby sets an agenda for future research.
13. For our purposes, we can treat claims about the nature of an ecosystem, claims about the identity of an ecosystem, and claims about how we identify ecosystems as interchangeable. Important distinctions can be made among these locutions, but they do not bear on our main points.
14. These heuristics are justified in the same way as more formal criteria for degradation, and they reflect *our* best estimate of the kinds of heuristics that are

justified by a critical assessment of current evidence. As indicated above, they are culturally loaded, and not everyone will initially accept them. We believe, however, that a process of critical reflection in which biases that might be related to self-interest are questioned would lead most people to adopt such defeasible heuristics.

15. We are indebted to Ned Hettinger for drawing this point to our attention.
16. Notice that this is analogous to intuitively plausible claims about a natural disaster being bad for human communities because it harms many community members and disrupts processes on which the functioning of the community depends. Such a disaster may violate the autonomy of a community without involving any domination.
17. James Griffin, *Value Judgment* (New York: Oxford University Press, 1996); Geoffrey Sayre-McCord, "Coherentist Epistemology and Moral Theory," in Walter Sinnott-Armstrong and Mark Timmons, eds., *Moral Knowledge?* (New York: Oxford University Press, 1996). Paul Taylor uses a coherentist moral epistemology to justify his environmental ethic in *Respect for Nature* (Princeton, N.J.: Princeton University Press, 1986).
18. Dale Jameison talks of the presumptive wrong in depriving animals of liberty in "Zoos Revisited," in Bryan Norton, Michael Hutchins, Elizabeth F. Stevens, and Terry L. Maple, eds., *Ethics on the Ark* (Washington: Smithsonian Institution Press, 1995).
19. A common worry about extensionist arguments of the above sort is that they create a slippery slope that leads to the moral value of machines, which is taken to be a reductio ad absurdum of the arguments. It is tempting to respond by specifying the features that enable us to distinguish between those entities to which moral value can be extended and those to which it cannot. But such responses inevitably fail because very fine grained continua can be invented between any kinds of entities, making stopping points seem arbitrary, and because claims that the distinguishing features are morally relevant differences will appear question begging when taken by themselves. We believe that pragmatic considerations such as the practicality of living with an ethical extension can serve to justify the regions where we stop our progress down the slippery slope. Such considerations suggest we need to develop an ethic pertaining to nonhuman organisms and their communities; we do not need an ethic governing our behavior toward machines, not yet anyway.
20. We also talk of actions of groups and the responsibilities groups must take for their actions. There may be philosophical reasons for resisting talk of collective actions and responsibility, but this is not the place to rehearse arguments for and against different forms of holism. Our point is that the extension of autonomy to groups does not require great feats of imagination; rather, it seems to many people to be a natural way to express certain points of view. This is *not* an argument that groups are autonomous but rather one piece of defeasible evidence that any theory about groups would need to address.
21. Taylor, *Respect for Nature.*

22. See the following references for positive visions of CFA and critiques of the movement to industrial agriculture: Wendell Berry, *The Unsettling of America* (San Francisco: Sierra Club Books, 1977); Gene Logsdon, *At Nature's Pace: Farming and the American Dream* (New York: Pantheon, 1994); Brian Donahue, *Reclaiming the Commons* (New Haven: Yale University Press, 1999); and Wes Jackson, *New Roots for Agriculture* (San Francisco: Friends of the Earth, 1980).
23. See R.V. O'Neil, Donald Lee Deangelis, and Garland E. Allen, *A Hierarchical Concept of Ecosystems* (Princeton, N.J.: Princeton University Press, 1986).
24. Stephen J. Gould, "The Golden Rule—A Proper Scale for Our Environmental Crisis," *Natural History* 99 (September 1990): 24–30.
25. We are grateful to Ned Hettinger, Thom Heyd, and Jon Jensen for their many insightful comments on earlier drafts of this paper.

PART III

Management, Restoration, and the Autonomy of Nature

A PARADOX?

SEVEN

Homo Administrator: Managing a Needy Nature?

DEAN BAVINGTON

THOSE OF us interested in environmental issues, whether or not we call ourselves environmentalists, are presented with increasing evidence of a global environment in need of our help.[1] We are told that nature, once conceptualized by Western culture as a bountiful enemy to be aggressively subdued and defended against, has been conquered, tamed, predicted, and controlled. The wild frontier has been captured, in many instances destroyed, and has been replaced with monocultures or domesticated farms whose sole purpose is to service the utilitarian interests of human beings. Those areas that have not been destroyed or domesticated are increasingly described as suffering from fragmentation and encroachment.

Furthermore, the monitored, managed, tamed, and domesticated areas of nature once thought to be both predictable and easily controlled are proving to be chaotic and complex, as evidenced by the failure of numerous management schemes (the collapse of the Newfoundland cod fisheries in 1992 provides a classic example of this trend). In our current context, the predominant view of nature is that of a sick patient in need of care. From this understanding of nature, we see the emergence of analytical frameworks such as ecosystem health, resource management, endangered species conservation, and environmental restoration.

Management strategies flourish in this environment and are offered as the solutions to the problems created by supposedly unenlightened exploitation and unknowledgeable, outdated husbandry. The metaphors, which permeate the popular imagination, focus on the belief in limits, scarcity, and depletion: people claim they are crewmembers on "spaceship earth." The spaceship metaphor leads us to the conclusion that our planet is in need of constant managed servicing if we are to avoid a fatal malfunction and breakdown. It becomes the responsibility of *Homo administrator* to fix, optimize, and contend with the functions of the biosphere,[2] a biosphere now described as a giant cybernetic ecosystem driven by energy flows and nutrient cycles. It is proclaimed that management of all individuals, populations, communities, watersheds, and ecosystems on the planet is necessary; they are all in need of help because of direct and indirect human disturbances.

Management is proposed not only to avoid the extinction of *Homo sapiens* and to ensure a continuous provision of goods and services, but specifically in order to protect all life on the planet from what E. O. Wilson (1992) has called the sixth, and only anthropogenic, mass extinction. This global management project in service of a needy nature challenges resource managers like never before. It claims to be motivated by a willingness and desire to help and to allow for both greater efficiency of the system being managed, and increased survival value for those lucky enough to be targeted for administration. I would like to propose, however, that such management may in fact deliver disabling "help": increased control over humans and nonhumans and their surroundings and the eradication of autonomous coping and self-organizing ability.

Global management proponents are fond of reminding us that we are involved in an environmental crisis: the survival of the planet is at stake, and so this crisis serves to justify drastic interventions to ensure our survival. In this context of apocalyptic crisis, managerial thought dominates ecosocial imaginations constraining the type of world that can be imagined and the patterns of relationships that become normalized between nature and ourselves. It is with a critical eye toward this framework of managerial thought, and the relationships it fosters, that I will discuss the emerging iatrogenic effects accompanying attempts to service the growing expanse of "needy nature."

The term "iatrogenic" is usually associated with medical terminology: it refers to illness that is produced through medical examination or treatment. Ivan Illich and John McKnight have extended its use to include all modern service interventions, such as social services, educational systems, and the criminal justice system. Illich and McKnight claim that these "helping" systems actually disable individuals and communities by building reliance on institutions and their associated experts and managers and that they thrive therefore on needs rather than capacities. Illich has explored these issues in numerous books including *Medical Nemesis* (1976), *Deschooling Society* (1971), *Toward a History of Needs* (1977), and *Tools for Conviviality* (1973). John McKnight (1995) has written about the iatrogenic effects that social services delivered by modern welfare states have on targeted communities. He has observed that in numerous cases the helping professionals delivering these services are disabling rather than enabling, and this outcome ensues as a result of the delivery of counterfeit care in the form of commodified and managed services aimed at individualized needs.

This paper expands the use of the term "iatrogenic" to include environmental management services such as environmental impact assessments, fisheries and forestry management, endangered species protection, park and reserve management, and other conservation activities that, in the current context of emphasis on economic growth, have a tendency to become iatrogenic, that is, to produce what they are designed to mitigate or prevent. To begin to explore this radical claim, I will outline Illich and McKnight's basic arguments, starting with their critique of service-based economies and the various types and stages of iatrogenesis. I will then discuss the connection among functional views of nature, disturbance ecology, and the management of a need-filled nature supposedly devoid of the capacity for autonomous self-organization, coping, and adaptation. The paper will conclude with a discussion of the antagonistic relationship that exists between management and autonomy and call for radical alternatives to managerial relationships between and among humans and nonhuman nature.

The Need for Need and the Service-Based Economy

Just as General Motors needs steel, a service economy needs deficiency, human problems, and needs if it is to grow. This economic need for need creates a demand for redefining conditions as deficiencies (McKnight 1995:29). John McKnight argues that human communities are undermined through service intervention. He suggests that the so-called helping and caring professions use the offer of their services to create dependent clients out of autonomous citizens. Flowing from arguments put forward by Ivan Illich (1971, 1973, 1976, 1977), McKnight argues that helping and caring professions are in fact a form of disabling help that undermines the very processes (interactions between autonomous and active citizens) that sustain healthy communities. Institutions and market interests promote clienthood and dependency, McKnight believes, and this leads to the devaluing of citizen action in favor of passive, expert service delivery. What is produced through these expert processes is a system that needs expanding needs to be discovered in increasingly dependent consumer-clients. McKnight illustrates how the typical foes against which we have waged countless losing wars—poverty, sickness, disease, and drugs—are not the problem; rather, the real enemies are "*a set of interests that need dependency masked by service*" (99, emphasis added). In other words, the more likely peril to a community's health is a health care system reliant on illness, social service providers whose existence depends on impoverished communities, or a criminal justice system that subsists on communities with a drug problem. In this way, McKnight radically reorients our thinking, turning problem solvers into the problem.

McKnight's critique was inspired by the work of Illich. In, *Medical Nemesis* (1976), Illich described three types of iatrogenesis or counterproductivities that he claimed perverted health care, transforming it into a "sick making enterprise" (7). The first of these, *clinical iatrogenesis* or *technical counterproductivity*, refers to the side effects accompanying manipulative interventions into the bodies of patients, such as the side effects that accompany drug therapies or doctor-induced death during an operation. Illich argued that, beyond a certain threshold, technical devices and interventions come to frustrate their stated aims and objectives, resulting in the production of damages. Managers and economists usually address this form of iatrogenesis by calculating the costs or risks associated with specific interventions, technical devices, and systems. It is important to note, however, that the negative consequences of iatrogenic effects are of a different kind from quantifiable economic costs. The patients who die from surgery or drug side effects represent damages and not simply negative costs reducible to a quantitative value (Farage and Samuel 2003).

Social iatrogenesis or *structural counterproductivity* was the second form of iatrogenesis identified by Illich. Structural counterproductivity occurs when the environmental conditions that permit individuals, families, and neighborhoods to control themselves and their social milieu disintegrate. What people once could do for themselves becomes something they must rely on experts to supply. This creates what Illich called a "radical monopoly" for the experts, who now are relied on to deliver what people once could do for themselves. At advanced stages of social iatrogenesis, the social environment (i.e., local communities, families, or neighborhoods) becomes understood as a threat to be protected against through external management as opposed to an autonomous enabling and caring environment. Social scientists have only recently begun to recognize this form of counterproductivity and attempt to grasp it through the capitalist (and somewhat vague) metaphor of social capital.

Illich termed his third, most pervasive, and yet least recognized form of counterproductivity *cultural iatrogenesis*. Making reference to the medical system, Illich argued that cultural iatrogenesis or *symbolic counterproductivity* "sets in when the medical enterprise saps the will of people to suffer their own reality" (Illich 1976:127). Symbolic counterproductivity refers to situations in which "autonomous coping" is thoroughly inundated by heteronomous management (external control). In these circumstances, "the ability of individuals to face their reality, to express their own values," and to perceive their world and their selves using culturally specific meanings, categories, and patterns of thought emerging from unique places and moments in time becomes increasingly frustrated and occluded (127–129). The threshold when cultural iatrogenesis sets in is much lower than that of technical or social iatrogenesis and is more difficult to recognize. Illich scholars Farage and Samuel observe, for example, that "people come to believe that walking is a means of transport, and an inferior one at that, much sooner than the environment is destroyed by car pollution. The deformation of the senses, whether of sight, sound or hearing occur [*sic*] much quicker than the deformation of the land by roads and railways" (2003:3).

Illich's observations on iatrogenesis and counterproductivities are based on a distinction between autonomous coping (i.e., self-governing) and heteronomous control or management (i.e., externally administered control).

> In *Homo sapiens*, "healthy" is an adjective that qualifies ethical and political actions. In part at least, the health of a population depends on the way in which political actions condition the milieu and create those circumstances that favor self-reliance, autonomy, and dignity for all, particularly the weaker. In consequence, health levels will be at their optimum when the environment brings out autonomous personal, responsible coping ability. Health levels can only decline when survival comes to depend beyond a certain point on the heteronomous (other-directed) regulation of the organism's homeostasis.
>
> (ILLICH 1976:7)

Illich argued that only political programs aimed at limiting the professional management of health would allow people to recover their autonomous coping powers for health care and that limiting professional management in all areas was integral to a "society wide criticism and restraint of the industrial mode of production" (10). While it has become dangerous in recent years to question environmental management in the context of deficit-cutting governments dedicated to economic globalization and market triumphalism, I believe a careful application of the ideas of McKnight and Illich is both worthwhile and urgently needed.

What insight can be gained by applying Illich and McKnight's critique of human service professions and iatrogenesis to environmental managers and ecoservice professions? Could it be that environmental management efforts to preserve, conserve, and restore nature in the context of ongoing global industrialization are magnifying the problems they set out to eliminate, leading to the emergence of technical, structural, and symbolic counterproductivities? Could it be that environmental management processes are iatrogenic and lead to the erosion of autonomous self-organization and coping in favor of heteronomous administration and managerial control? It seems plausible that a process similar to the one McKnight and Illich describe as operating in overindustrialized human communities is occurring in nonhuman communities that are increasingly being serviced by an emerging ecoservice sector comprised of environmental engineers, technocrats, ecoplanners, and other environmental management professionals. If this is the case, the discourse and techniques that describe humans as needy and expose them to iatrogenic servicing would increasingly be applied to nature with similar iatrogenic results.

Needs and the Functional View of Nature

Natural areas, in isolation from their human contexts, are increasingly portrayed, and accepted, as being in *need* of managerial interventions. A reductionist focus on endangered species and natural areas removes the emphasis from the surrounding social, cultural, economic, and political environments, and the philosophical assumptions that frame these areas, and places the deficiency in the endangered species or natural area itself as opposed to the broader context. Ironically, the very organizations that often perpetuate a "needy" view of nature are the ENGOs (Environmental Nongovernmental Organizations) that claim to be interested in saving nature for its beauty, intrinsic value, integrity, and stability. By focusing narrowly on a species or endangered area as an object of environmental management deficiencies, needs are located in the manageable object, and the stage is set for expert intervention into a passive and need-filled client.[3]

As Ivan Illich has noted with respect to Western society: "The historical movement of the West, under the flag of evolution/progress/growth/development, discovered and then prescribed needs. In this process, we can observe a transition from man [*sic*], the bungling toiler, to man, the needy addict" (Illich 1993:89). "Man the needy addict" has increasingly been joined by nature. Environmental management theory and practice have helped to shift the view of nature from a wild and untamed wilderness frontier (and later a tamed, controlled, and exploited nature) into a medicalized representation of nature as a sick patient in need of intensive care. This shift from wild to tame and then sick forms of nature is accompanied by the insertion of specific types of needs into nature. These constructions of nature encourage managerial relationships between humans and nature and support interests related to the servicing of nature's needs. In addition, the insertion of needs into nature is embedded within a thoroughly use-oriented or instrumental view of nature described as a set of inert natural resources, mechanistic functions, and cybernetic ecosystem services.

Under many environmental management frameworks, individual species, populations, communities, ecosystems, watersheds, and landscapes are reduced to the functional roles they perform in relation to anthropocentric desires and requirements. This functional view of nature allows for individual parts and processes to be identified and serviced as if they were part of a dynamic ecomachine. This functional mechanistic approach enables the discovery of missing parts, inefficient processes, suboptimal production, and redundancy in the system, which can be lost, consciously removed, or potentially enhanced. This approach has permitted environmental management to use the ideas of indicator species, minimum viable populations, and keystone species to prioritize "scientifically" which functional units are in need of help and which can be sacrificed to development in any particular area. When the function of a species in a particular ecosystem has been discovered and the necessary functions of that particular ecosystem have also been identified, then management priorities and plans can be set in place.

For example, if ten species perform a filtering function in a marsh, an environmental manager operating from a redundancy model could allow two of the species to go extinct and still maintain the filter function. If there is only one species to fulfill that function, however, the manager may become concerned and is likely to invest time and money toward protecting that component of the system. This functional approach downgrades the importance and value of individual species and reduces them to an instrumental function within a given system. It also reduces nature to a set of finite functions that ecomanagers assume can be understood, controlled, and manipulated while maintaining the dominant framing of ecological issues as reified scientific problems stripped from their complex ethical, socioeconomic, and political contexts. Perhaps most troubling is that the functional approach to nature serves to further undermine attempts to defend nature against development. After all, if we can design a biomachine that filters water better than natural kinds, for what do we need critters like brine shrimp or plants like cattails? This type of argument illustrates some of the problems and dangers of using a functional approach to describe nature and illustrates how easily functional models can lead to strong managerial approaches to human-nature relations that leave the idea of development and management uncritically examined.

It is readily apparent that instrumental arguments dominate environmental management approaches to nature. Ecomanagers are presented as all-knowing experts who can assess nature's needs and supply them while permitting development to continue either unimpeded or with minor adjustments. Most ecologists are not comfortable with the role of all-knowing environmental expert and are aware of their ignorance and the practical limitations on the application of their knowledge; however, ecomanagers frequently believe, and are forced to operate as if science and technology contain the answers to environmental problems. Preservation, conservation, and restoration issues are seldom simple; numerous ethical, socioeconomic, and political conflicts, as well as practical limitations on knowledge, exist in theory as well as in practice. The application of management science and technology occurs within the context of socioeconomic and political systems that structure our interactions with each other and the natural world. Strictly speaking, though, these larger contexts are unmanageable. To be addressed legitimately, they require long-term democratic deliberation and ethical judgments, not technically oriented administrative quick fixes.

Disturbance as Managerial Resource

I have discussed the functional view of nature to illustrate how the logic of environmental management has utilized the functional approach to ecosystems in order to map out deficiencies in nature and mine them by designing management plans that depend on managerial expertise to be successfully implemented. Other models within ecology can also lead to this outcome. As Wolfgang Sachs has pointed out,

> ecosystems theory, based on cybernetics as the science of engineering feedback mechanisms, represents anything but a break with the ominous Western tradition of increasing control over nature. How can a theory of regulation be separated from an interest in manipulation? After all, systems theory aims at control of the second order; it strives for controlling (self-) control. . . . Looking at nature in terms of self-regulating systems . . . implies either the intention to gauge nature's overload capacity or the aim of adjusting her feedback mechanisms through human intervention. Both strategies amount to completing Bacon's vision of dominating nature, albeit with the added pretension of manipulating her revenge.
>
> (SACHS 1993:32)

The needs mining discussed by McKnight and Illich with respect to professional services such as health care and education is also evident in other professions and is tied to discoveries of disturbance. The need for need, in order to feed growing service-based economies, emerges as a response to the discovery of disturbances in both human and nonhuman contexts. In both the social and natural sciences, interest has recently been placed on discoveries of continuous change, turbulence, disequilibrium, disturbance, and deficiencies in human and nonhuman environments. This is a break with past emphases on balance, homeostasis, and equilibrium in cultural and natural systems.

The ecological historian Donald Worster (1995) has traced the history of ecological ideas and has noted that present theories in the science of ecology tend to emphasize the role periodic and chaotic ecological disturbance and patch dynamics play in structuring landscapes, rather than studying the ordered progressive stages leading to climax community structure. He illustrates how ecologists and other natural and social scientists are presently fixated on disturbance and are seeing it everywhere. "Disturbance comes from a congeries of cultural and natural agents, including droughts, earthquakes, pests, viruses, corporate invasions, loss of markets, new inventions, crimes, federal laws, and even French literary theory. Disturbance *is* history. And a disturbed nature is a nature that has a history very similar to the history that humans make" (74).

With the increased discussions and discoveries of disturbance come calls for interventions to mimic, optimize, control, and fix these disturbed processes. For example, in human communities, if a tragedy occurs in a school, crisis management teams and trauma counselors are called in to provide professional psychological counseling for all the students and staff. If fire once naturally disturbed temperate forest ecosystems, then controlled burning or clear-cutting are required in order to maintain the natural condition by mimicking natural disturbance. Indeed, the government of British Columbia has noted that certain types of clear-cutting can help to maintain natural disturbance patterns (Forest Service, 1995). If a development project is to be proposed, a thorough environmental impact assessment is in order, not to stop the development but to mitigate abnormal disturbance to the natural environment. With all of these approaches, service providers are needed in the form of managers, professionals, and experts. To intervene legitimately, these service providers must present themselves as possessing valid and comprehensive instrumental knowledge about the behavior and dynamics of the systems they are targeting.

An article in the Canadian national newspaper the *Globe and Mail* describing the premeditated murder of an entire extended family that occurred in Vernon, British Columbia, illustrates the contemporary service response to a community disturbed by a crisis in Canada; it also illustrates how professional service interventions in disturbed communities have expansionist tendencies:

> Richard Wilford, program co-ordinator for the North Okanagan victim-assistant program, pulled together plenty of expert help: psychologists, social workers, registered counsellors, mental-health workers, youth and family workers, government social-service workers and 28 volunteers with training in trauma counselling and crisis intervention. The team's first job was to identify those who might be affected. They drew up a lengthy list that included members of the immediate family; neighbours who saw the shooting; family friends; pro-

> fessionals such as police, ambulance drivers and hospital workers, who dealt with the bodies; members of Vernon's Sikh community; the slain children's schoolmates; the adult victims' co-workers; ***the entire city of Vernon***; and local reporters who usually cover much tamer stories. . . . [When asked if this drastic action was creating victims, Richard Wilford said,] "We are not [creating victims]. . . . If these people do not talk it out, then it remains stuffed in and the next time they'll have two things to deal with. It's a public health issue. . . . ***We cannot predict who will suffer, so we inoculate everyone.***"
>
> ("THE VERNON MASSACRE" 1996; EMPHASIS ADDED).

The extent to which service provision aimed at individual clients has been visited on human communities is astounding. What is seemingly forgotten is that, not unlike all drug therapies, these service interventions are not neutral; they have side effects and can produce counterproductivities. Such intense service provision could in fact undermine communities by framing them as collections of needy individualized clients who should speak to paid professionals rather than exercising autonomous coping strategies together with their friends, neighbors, and fellow citizens. The level of professional service intervention is perhaps best read as an indication of the erosion of autonomy in Vernon, British Columbia, and other environments targeted for external intervention.

Examples of the expansive servicing of disturbed nonhuman environments are also available in the resource management literature. An article describing the need for interdisciplinary teams to manage the looming groundwater crisis reveals this tendency:

> Environmental management, to a greater extent than any other management area, must be multidisciplinary and interdisciplinary. Optimal groundwater management decisions should be made, or strongly influenced by, multidisciplinary teams. . . . Such a team might include persons educated in communications, ecology, economics, engineering, ethics, geology, hydrology, information systems, law, planning, politics, public administration, public health, public participation, sociology, and social psychology. Each of these participants should be familiar with the language and fundamental concepts of other disciplines represented, be able to communicate his or her insights to nonspecialists, and be experienced in multidisciplinary decision making. The team leader should be a capable environmental generalist as well as a trained and experienced facilitator. Support for this team should be provided by a multidisciplinary staff qualified to produce information that is suitable in form and content for the multidisciplinary team.
>
> (GOLDFARB 1991:131)

It is interesting to note that there is no mention of community members, citizens, or nature in the groundwater environmental management team. It is not incidental therefore that the panel in this example is looking into resource management issues: when managed, nature is reduced to a collection of inert natural resources, and citizens are reduced to passive clients to be administered by multidisciplinary management teams. With the combined emergence of a growing service-based economy and the increased discovery of and attention to disturbance, a growing number of conditions of both human and nonhuman nature are being converted into problems to be solved through professional service interventions, delivered by either public or private interests.

The move to a service-based economy brings with it a shift not only in the way we think about human beings but also in the way we conceptualize nature. Increasingly, polluted areas, endangered species, endangered habitat, the maintenance of minimum viable populations, the monitoring of species at risk, the managing of parks and precious biological reserves, and various biological inventory projects have created a growing environmental management industry. We can expect that the servicing of these newly discovered natural resources and ecological processes will develop according to the logic of late capitalism and the global marketplace. In order to create and feed a growing economy, the servicing of a need-filled nature will gain importance. Under these circumstances, the potential for or existence of autonomy in what is serviced is dramatically reduced, and the stage is set for multiple forms of iatrogenesis or counterproductivity.

Management and the Autonomy of Nature

The etymology of the word "management" is illustrative of the antagonistic relationship management has with the idea of freedom and autonomy. "Management" entered the English language in the sixteenth century from the Italian *maneggiare*, whose Latin root is *manus*, meaning "hand" (OED, 1995). "To manage" meant to train, handle, and direct the movements of a horse, to break its wildness, and to administer it through its paces. In the context of environmentalism, it is interesting to note that management has its roots in the domestication of a wild animal with the associated notions of taking charge and controlling wildness (Bavington 2002).

The object of management in the sixteenth century was restricted to the training of horses to gallop, trot, and high step by controlling and regulating the horses' separate movements and gait (Samuel 1995). Since the sixteenth century, however, the application of management has expanded exponentially; it now encompasses increasing areas of human and nonhuman life while maintaining its original association with controlling, directing, training, and administering. The highly successful scientific management movement institutionalized by Frederick Taylor in the early nineteen hundreds was focused on quantifying and rationalizing the movements and timing of factory workers so they could be more easily controlled and trained to be more efficient and productive (Taylor 1911). Management implies a steering procedure, a controlling and directing presence over animals, people, processes, and things. "Management is based on the existence of a symbolic system that corresponds to, but greatly simplifies, some 'real' system, which can thereby be brought under rational control" (Holm 1996:179).

If, as Eric Katz notes in this volume, autonomy is the opposite of domination, the "control," "training," and "handling" meanings at the root of environmental management deserve to be scrutinized seriously. This is especially true when management is promoted as *the* solution to ecological problems. As Alasdair MacIntyre observes, to the extent that management is effective, it is manipulative and therefore is an activity with moral and political significance (Knight 1998). Management is not a neutral, value-free tool. Rather, management deploys instrumental rationality to find the most efficient means to achieve a given end. Management implies a hierarchical two-way coercive relationship between the manager and the managed. The freedom of individuals, populations, and ecological processes to evolve, adapt, and develop autonomously is severely restricted with the application of management. Indeed, for management, freedom and autonomy in human and nonhuman contexts always represent problems in need of effective control, handling, and training. "Others are to be the passive recipients of what they as managers effect. This hierarchical division between managers and managed is thus legitimated by the superior knowledge imputed to themselves by the managing reformers who have cast themselves in the role of educator" (MacIntyre, quoted in Knight 1998:231).

A central problem with the management model from the perspective of autonomy is that it is based on deficiency and the desire to break and remake targets of management. Only after a managed object has been broken of its wildness, simplified, tamed, and made predictable can it be instrumentally used and relied on to self-manage. This self-management in the wake of effective external control is not equivalent to premanaged autonomous coping, adapting, and self-organizing that is internally oriented (Knight 1998). When tied to economic growth in the needs-servicing sector, management relies on the identification of an expanding array of deficiencies in the individuals and environments serviced by *Homo administrator*. This fundamental conflict illustrates why the needs-servicing economy may itself be subverting both human and nonhuman autonomy. Instead of concentrating time, money, and resources on the protection and expansion of autonomous capacities, identifying underlying structural forces that drive the creation of needs, and exploring and supporting activities that minimize the need for management and administrative servicing of human and nonhuman environments at the first instance, attention is focused on creating more effective managerial techniques within an increasing crisis-prone status quo.

Conclusion

A trick is played on those of us who care for nature when management interventions are presented as liberating solutions for the highly complex and problematic ecosocial issues surrounding human-nature interaction. Environmental management promotes itself as a practical solution to environmental problems, claiming to implement systems of crisis intervention increasingly in the name of planetary survival. As the biologist Gary Meffe notes, "The ultimate outcome of this techno-arrogance is the increasingly intensive and essentially perpetual management of a multitude of species in a world unfit for their natural existence" (1992:354).

There has been much talk in resource management of moving away from expert-oriented, top-down "command and control" approaches toward nature and the promotion of stakeholder-driven "adaptive management" with a coping rather than controlling stance on the part of managers. Many of these approaches advocate a shift of managerial control away from nature and onto human beings (Bavington 2002). These new modalities of resource, environmental, and ecosystem management do represent a change in the identity and orientation of the manager and recognize the existence and importance of autonomous coping and self-organizing processes in nature and the arrogance of a belief in our ability to control complex ecological systems completely. However, without a fundamental questioning of the idea of management that was deployed to generate industrial growth, these moves lead to entrenchment and expansion of the managerial status quo, applying subtler, yet more effective, manipulative means rather than a fundamental change in human-nature relations (Adams 2003). Resource and environmental management continues to be largely focused on the *efficacy* of management (linking various technical means to predetermined ends that value human and nonhuman nature instrumentally) as opposed to the moral and political *legitimacy* of *Homo administrator* and the possibility of something altogether different.

By framing the ecocrisis as a problem amenable to managerial solutions, deeper questions surrounding how we ought to be living and how we would like to live in the future are obscured by shallow attempts to survive in the context of the crisis-ridden ecological and socioeconomic status quo. This outcome is illustrated in the historic pattern of fisheries management in Newfoundland and Labrador as well as numerous other examples of resource management failure across Canada and around the world.

What is most troubling is the fact that the ecocrisis is being managed! We continue to accept and seem largely blind to the many types of iatrogenesis associated with managerial human-nature relations. The capitalist economy of service provision relies on the deepening of the ecocrisis for its growth and prosperity. Because effective management relies on heteronomous control, it continues to represent a threat to autonomous coping and self-organization capacity in human and nonhuman individuals, communities, and ecosystems. When nature is constructed as a needy client, its autonomy evaporates as quickly as when it is cast as a passive storehouse for raw materials. Granting agency in the form of client and consumer identities leaves no room for autonomy. At best, client and consumer identities can lead one to ask why an identity as an autonomous member of a citizen-based community is not possible and may lead to questioning hierarchal and manipulative relationships. Aldo Leopold's land ethic continues to raise such troubling questions for resource managers vis-à-vis nature (Leopold 1949).

Management places severe limits on our collective ecosocial imaginations, directing attention to the institutions of nation-states and private corporations who are the leaders in managerial practice while deflecting attention away from place-based communities embedded in particular ecosocial histories, with autonomous coping and adaptive capacities as well as challenges to enhancing and maintaining that autonomy.

Far from removing our attention from the very real ecological disasters and environmental crises that we are presented with daily, I hope that this paper plays a role in questioning the dominant pattern of thought with which we are approaching environmental issues. Most cultures embed the attempt to bring dynamic order to human-nature relations within a larger sociocultural matrix, but modern Western societies' approaches to ordering human-nature relations have been predominantly tied to the controlling mechanisms of management located in state and market institutions. Rather than continuing to model human-human and human-nature relations on the control, directing, and training associated with breaking wild horses, I propose that we begin to renew spaces for alternative frameworks of thought and action that move beyond management, in favor of protecting and encouraging ecological and socially autonomous coping and self-organization.

NOTES

1. An earlier version of this paper has been published as *The Iatrogenic Effects of Environmental Management: Servicing a Needy Nature?* Faculty of Environmental Studies Occasional Papers 4, no. 1 (York University, Ontario, Canada, September 1998).
2. Literally, we have "Man the Manager" created under this framework. I wish to thank Lee Herin for his help in constructing this particular phrase.
3. It is important to note at this point that these observations are not meant to dismiss the severity of the assaults that are occurring on nonhuman individuals, natural communities, and environments, just as Illich and McKnight are not to be read as dismissing the very real problems of education, health care, crime prevention, and poverty. Illich and McKnight question the legitimacy of current ways of thinking about these problems and the institutional, professional, and managerial solutions mounted to fix them. The goal of this paper is to question the legitimacy of environmental management as an appropriate response to a supposedly needy nature.

REFERENCES

Adams, W. 2003. When Nature Won't Stay Still: Conservation, Equilibrium and Control. In *Decolonizing Nature: Strategies for Conservation in a Post-colonial Era*. Eds. W. Adams and M. Mulligan. Earthscan: London.

Bavington, D. 2002. Managerial Ecology and Its Discontents: Exploring the Complexities of Control, Careful Use and Coping In Resource and Environmental Management. *Environments* 30 (3): 3–21.

Farage, S. and Samuel, S. 2003. *Modern Rituals and Myths: An Analysis of Transportation as an Example of Services*. Presentation to the Ivan Illich Center for Intercultural Documentation, Lucca, Italy.

Forest Service, B.C. Ministry of the Environment. 1995. *Forest Practices Code of British Columbia: Biodiversity Guidebook*. Government of British Columbia: Victoria, British Columbia.

Goldfarb, William. 1991. Groundwater: The Buried Life. In *Ecology, Economics, Ethics: The Broken Circle*. Eds. Herbert Bormann and Stephen Kellert. Yale University Press: New Haven.

Holm, P. 1996. Fisheries Management and the Domestication of Nature. *Sociologia Ruralis* 36 (2): 177–188

Illich, Ivan. 1971. *Deschooling Society*. Heyday: Berkeley, California.

———. 1973. *Tools for Conviviality*. Heyday: Berkeley, California.

———. 1976. *Medical Nemesis*. Heyday: Berkeley, California.

———. 1977. *Toward a History of Needs*. Heyday: Berkeley, California.

——. 1993. Needs. In *The Development Dictionary*. Ed. Wolfgang Sachs. Zed: Halifax, Nova Scotia.

Knight, K., ed. 1998. *The Macintyre Reader*. University of Notre Dame Press: Notre Dame, Indiana.

Leopold, A. 1949. *A Sand County Almanac*. Oxford University Press: New York.

McKnight, John. 1995. *The Careless Society: Community and Its Counterfeits*. Basic: New York.

Meffe, G. 1992. Techo-Arrogance and Halfway Technologies: Salmon Hatcheries on the Pacific Coast of North America. *Conservation Biology* 6 (3): 350–354.

Oxford English Dictionary. 1995. Oxford University Press: London.

Sachs, Wolfgang. 1993. Environment. In *The Development Dictionary*. Ed. Wolfgang Sachs. Zed: Halifax, Nova Scotia.

Samuel, Sajay. 1995. *Of Horses and Men: Managerialism at the End of the Millennium*. Phi Beta Kappa Speech (unpublished). Pennsylvania State: State College.

Taylor, Frederick. 1911. *The Principles of Scientific Management*. Harper and Brothers: New York.

"The Vernon Massacre." 1996. *Globe and Mail*, April 13, D5.

Wilson, Edward. 1992. *The Diversity of Life*. Norton: New York.

Worster, Donald. 1995. Nature and the Disorder of History. In *Reinventing Nature? Responses to Postmodern Deconstruction*. Eds. Michael Soule and Gary Lease. Island: Washington, D.C.

EIGHT

Purple Loosestrife and the "Bounding" of Nature in North American Wetlands

JOHN SANDLOS

ONE OF the better-known themes among the Christian parables is the separation of the righteous and pure from the impure and the unholy.[1] In one notable example, Christ recounts the experience of a farmer whose enemies have come in the night to plant weeds in his wheat field. When one of the servants asks whether he should pull the weeds, the farmer responds: "No; lest in gathering the weeds you root up the wheat along with them. Let both grow together until the harvest; and at harvest time I will tell the reapers, Gather the weeds first and bind them in bundles to be burned, but gather the wheat into my barn" (Matthew 13:29–30). While this parable may appear to the modern rational mind as nothing more than a quaint expression of Christian ethno/agroscience, the persistent reader is rewarded with a deeper understanding of the importance of weed control further down the page. At the request of his disciples, Christ explains the parable in terms of a larger cosmological purpose: "the weeds are the sons of the evil one, and the enemy who sowed them is the devil; . . . As the weeds are gathered and burned with fire, so will it be at the close of the age. The Son of man will send out his angels, and they will gather out of his kingdom all causes of sin and all evildoers. . . . Then the righteous will shine like the sun in the kingdom of their Father" (Matthew 13:39–43).

Clearly, to the early Christian, the act of weeding carries a greater symbolic weight than is implied by the basic physical and ecological needs to produce an adequate food supply. The obvious lesson of the parable—how and when to remove weeds—invites the reader to participate in the larger cosmic struggle of establishing God's kingdom on earth. The physical source of the symbol and the symbolic action are not separate, thus allowing the true believer to participate fully in the established natural and moral order of the universe. Valuative ethical principles and the act of cultivating nature are not separate in this worldview; the farmer's field is clearly understood as a social and cultural space (i.e., the emerging kingdom of God) throughout the parable.

It is quite likely that the denizens of contemporary agricultural and botanical institutions devoted to the loosely defined practice of weed science no longer see themselves as participating in a cosmic struggle to establish God's kingdom on earth. Indeed, what constitutes the traditional scientific worldview is a life world that is devoid of purpose, mystery, and moral significance. The Cartesian universe only allows for a single actor—the knower—to approach a natural world that waits passively to be known and thus can no longer be a repository for the sociocultural residues of myth-making humanity. Indeed, the "natural" knowledge derived from the scientific practice of weed control exists only for a single purpose: to control and eradicate the plants that humans designate as weeds.

Science, however, can never be divorced from its complex sociocultural and political origins. Even the most seemingly innocuous and innocent disciplines within the scientific enterprise (even weed science) produce and reflect a social worldview as much as they merely record the objective facts of nature. What follows is a discussion of the scientific work surrounding the European "weed" purple loosestrife (*Lythrum salicaria*), considered invasive to North America. The intention is not to prove or disprove the scientific orthodoxy concerning the ecological effects of this aquatic perennial. Rather, my goal is to illustrate that the science surrounding purple loosestrife has neither divorced itself from social influence nor ceased to act as an arbiter in social and political affairs. Given this expanded context, the scientific war against purple loosestrife represents not merely an effort to liberate nature from the depredations of a noxious weed but part of the larger mythic battle for control over the universe.

The Purple Loosestrife Problem

An information pamphlet released in the mid-1990s by Ducks Unlimited describes the threat posed to North Americans by expanding populations of purple loosestrife. Under the ominous "The Invader," there is an account of the spreading scourge: "Purple Loosestrife invades wetlands and gradually takes them over. The weed ultimately chokes out all native vegetation, creating a dense purple landscape almost totally devoid of wildlife. Purple loosestrife came from Europe over a century ago. Its unrelenting spread across North America was aided by the absence of natural predators" (Ducks Unlimited n.d.).

While the quoted material obviously employs the language and tone of a popular information leaflet, it does effectively summarize the main arguments the scientific community has put forward in support of controlling purple loosestrife: native wildlife doesn't use it as habitat or food, native wetland plants are displaced by the invader, and the spread of the plant is facilitated by a lack of natural predators in North America.

Much of the evidence for these assertions comes from a 1987 paper authored by Daniel Q. Thompson, Ronald L. Stuckey, and Edith B. Thompson. A remarkable document in many ways, the paper recounts the history of purple loosestrife colonization in North America, describes its apparent impacts, and discusses several possible control methods. It includes a synthesis of a vast body of purple loosestrife research, as well as the field observations of the authors. It has become, in many ways, the bible of purple loosestrife management and control.

Much of the evidence regarding the ecological impacts of purple loosestrife found in the Thompson paper is, however, highly questionable (the authors acknowledge from the outset that quantitative evidence for their assertions is lacking). The authors cite as evidence a series of case studies derived from the management records and personal observations of several artificially created waterfowl impoundments in central New York State (one is tempted to speculate whether the initial flooding of the impoundments created ideal disturbed soil sites for purple loosestrife colonization, as is the case with irrigation projects). The cover of Thompson's paper also features "before" and "after" photographs that depict a transition from a green to purple landscape over a period of ten years. The first photograph, of the supposedly healthy wetland, was, however, taken in June, well before the flowering season, while the second photograph was taken in August, during the height of the loosestrife flowering season, thus making comparison of the two photographs extremely tendentious. A recent paper by Mark Anderson (1995) has suggested that the largely visual nature of Thompson's evidence constitutes a subjective interpretation of qualitative changes in the wetland biotic community.

The reliance on qualitative visual evidence in Thompson's analysis raises doubts as to the validity of the conclusions. It is not clear from the evidence presented whether purple loosestrife actually crowds out native species of plants because there is no evidence supporting a decrease in the biomass of other plant species (Anderson 1995). It is possible that the biomass of other plant species remained constant or increased much more slowly in relation to the increasing biomass of purple loosestrife (Anderson 1995). Anderson cites his own observations and several from other sources that suggest competition between plant species can occur in stands of purple loosestrife. According to Anderson, the apparent increase of purple loosestrife density in relation to other plants is actually due to an increase in the percent cover of the plants as they age and mature rather than an increase in the number of genets in a given stand.[2] His research found no correlation between the density or percent cover of loosestrife and the species richness of a given area (Anderson 1995). Similarly, a survey of Wellington County in southwestern Ontario, Canada, found 115 sites with purple loosestrife present but reported that "the majority of plants were scattered over an area or in clumps with only 11 solid stands of loosestrife" (Federation of Ontario Naturalists 1992). A concurrent survey in nearby Lanark, Ontario, cited local testimony from near the Leeds county line that suggested the area had once been "carpeted" with purple loosestrife. There was no loosestrife found in the area during the 1992 survey (Federation of Ontario Naturalists 1992).

Thompson's assertion that loosestrife is not utilized by North American fauna also deserves some consideration. Anderson cites reports and observation of the use of purple loosestrife as a source of nectar and pollen by fourteen separate species of insects and as a source of food by white-tailed deer, muskrat, and rabbits. He has also observed American coots, pied-billed grebes, black-crowned night herons, American goldfinches and gray catbirds nesting in stands of loosestrife. Red-winged blackbirds are known to nest preferentially in stands of loosestrife (Keddy 1992).

While the anecdotal nature of Anderson's evidence proves nothing regarding the interaction of purple loosestrife with native flora and fauna, it does call into question Thompson's claim to have derived the absolute truth concerning native floral and faunal interactions with purple loosestrife from objective scientific facts. Indeed, the use of subjective judgments in Thompson's work is most readily confirmed by his remarkable statement that "although we need quantitative measurements of the effects of various stages of *L. salicaria* invasion on the structure, function, and productivity of North American wetland habitats, the replacement of a native wetland plant community by a monospecific stand of an exotic weed *does not require a refined assessment* to demonstrate that a local ecological disaster has occurred" (1987: 25; my emphasis).

Despite the obvious knowledge gap in Thompson's work, the literature published on purple loosestrife since 1987 has largely dealt with control of the plant rather than a further examination of ecological interactions with native flora and fauna. Of the literature surveyed, a total of fifteen papers used Thompson's paper as proof or partial proof that purple loosestrife degrades wetland areas as wildlife habitat. Several other papers listed in the bibliography did not make reference to the ecological impacts of purple loosestrife. The urgency these papers outline for various types of control programs suggests the authors accept the apparent negative ecological impact of purple loosestrife as a given fact.[3] As Heather Hager and Karen McCoy (1998) suggest in their own recent review of the relevant literature, "a major environmental effort is involved in the establishment of successful biological control for purple loosestrife. However, the fundamental questions still have not been answered: *what are the effects of purple loosestrife on native wetland flora and fauna and are they significant*? This is not to suggests that purple loosestrife does or does not have detrimental ecological effects, only that there has been inadequate research to answer the question" (1073).

Scientists Knowing Nature

> *Nature has no system; she has—she is—life and development from an unknown center to an unknowable periphery*
>
> —GOETHE

While the scientific criticisms offered by Anderson and Hager are well taken, the uncritical acceptance of Thompson's work is not merely a reflection of inadequate data or quantitative analysis; it illustrates a deeper problem with the scientific claim of objectivity. Despite the best attempts of the "objective" scientist to divorce fact from value, he/she must construct disembodied facts into a core of meaning that forms a conclusive narrative structure for the scientific text. Like a historian, the scientist never simply records objective facts but instead arranges a body of evidence into a structured account that attempts to produce a definitive meaning. The process of meaning production places the scientist and his/her readers in a discursive community that is intimately related to the larger process of cultural reproduction. In other words, the creation of meaning can never be divorced from the social and ideological context in which it is embedded. As Haraway puts it, "what determines a 'good' story in the natural and social sciences is partly decided by available social visions of these possible worlds. Descriptions are determined by vision; facts and vision are perceived through stories; the worlds for which human beings contest are made of meanings" (1988:80).

The argument here is that science *necessarily* is value-laden because it is partly a product of, and partly produces, the cultural and ideological context in which it immersed. There are, after all, countless examples of scientific paradigms that appear, by virtue of hindsight, to be derived from the prevailing cultural metaphors of a given era. Donald Worster argues in *Nature's Economy* that the Newtonian model of a "clockwork" nature emerged within a late-seventeenth-century society that was slowly being transformed by temporal regimentation and machine production, while Darwin's industrial capitalist society of progress and heightened economic competition was both reflected and justified by his theory of natural selection. Worster further contends that Frederic Clements's equilibrium and climax community models of plant communities (which became the dominant paradigm in ecology for decades after its emergence in the 1930s) were firmly rooted in the Keynesian notion of steady-state economics, while more recent chaos models of ecology (i.e., disturbance; fire ecology) that have emerged in recent decades have adopted many of the cultural metaphors associated with the crash-and-burn logic of late capitalism. Such divergent views suggest that ecological science, far from being the most natural of all the sciences, has not yet been able to isolate and explain the nature of nature.

One need not, however, focus only on the larger paradigmatic examples to detect the presence of ideology within the natural sciences. As the geneticist Richard Lewontin notes, even the most mundane practitioners of science are subject to social influence in their work: "Most of the ideological influence from society that permeates science is a great deal more subtle [than the Darwin example]. It comes in the form of basic assumptions of which the scientists themselves are usually not aware, yet which, in turn, serve to reinforce the social attitudes that gave rise to these assumptions in the first place" (Lewontin 1991:10).[4]

Lewontin's analysis at least suggests that the problem with the science of purple loosestrife is located not with mistakes made in the field but instead in the refusal of the scientist to recognize and make explicit the social and cultural boundaries and categories that encapsulate the story—or the process of meaning production—associated with purple loosestrife in North America. In spite of the lack of definitive evidence to prove the adverse ecological impacts of the plant, scientists feel empowered to speak and (more important) to act authoritatively on behalf of nature (i.e., wetlands, native species, etc.). In the absence of any requirement to question the basic political and cultural assumptions associated with particular scientific disciplines (in the case of weed science, the basic cultural assumptions required to designate certain plants as noxious), the scientist maintains status as a powerful actor and arbiter in the unfolding drama of humanity's relationship to nature. Whether a particular body of scientific knowledge is produced at a university, a government agency, or a private organization, it is often accorded the status of a privileged truth, a process of knowledge production that provides the necessary social legitimacy to the intense management and manipulation of the natural world.

Certainly, there are many exceptions to this general pattern, and science is not necessarily monolithic in its intellectual orientation. Radical philosophical criticisms of the absolute truth claims of management-oriented science have emerged from within the work of such rogue scientific practitioners as

Lewontin and Thomas Kuhn (1962). Moreover, disciplines such as contemporary theoretical physics are often devoted to pure research and have adopted uncertainty and even a certain degree of mystic wonder as important intellectual principles (see Zajonc 1993; Gleick 1987). In his *Nature's Economy* (1994), Worster argues that two distinct traditions in environmental science—the imperial tradition, associated with the "domination of nature" school of Bacon and Newton, and the less interventionist Arcadian school tied to such naturalists as Gilbert White and romantics such as Thoreau—have existed from the very beginnings of the discipline in the eighteenth century. One could argue that the Arcadian/imperial distinction has survived as a contemporary phenomenon, the former in the preservationist "crisis discipline" of conservation biology and the benevolent, community-oriented restoration ecology advocated by Andrew Light in this volume and the latter in the intensive, production-oriented scientific management of natural resources favored by government and industry since its widespread application throughout North America in the Progressive era (see Worster 1994:261–271; see also Hays 1959).

Nevertheless (and despite the importance of the countertraditional intellectual movements in science), the intellectual hegemony of the imperial school is so pervasive that the status of scientific manager as an arbiter of absolute truth remains a prerequisite to official intervention in the natural world. The primary task of the scientific manager is thus not merely to study the natural world in all its variety and permutations but to propose technical solutions to problems such that the ends are justified simultaneously with the means and the marriage of empiricism and the management bureaucracy is complete (see Bavington in this volume).

As a product of this scientific management discourse, the vast majority of the recent purple loosestrife literature deals with the how? of controlling the plant rather than the philosophical why? of restoring the altered landscape. According to the ecologist Walter Westman, this lack of a why? perspective reveals "the *subjective judgments* required to implement an exotic species removal program," particularly when the larger questions concerning invasive species remain unanswered: "These questions exemplify troublesome questions in current policy: the arbitrariness of designating the time of European contact, since it is only one point on a continuum of escalating human influence on biotic dispersal . . . the difficulty of defending the genetic purity of local strains on hypothetical grounds. Beyond these cases looms a larger issue: if the process of invasion elicits adaptive responses, by what criteria shall the presence of an exotic be judged harmful?" (1990:253).

The mere absence of the why? questions in the discussion of purple loosestrife does not, however, mean that there is an absence of motivation for the control programs. The social matrix of social and ideological meaning production associated with the scientific effort to restore life to the (supposedly) dead purple landscape of a loosestrife-infested wetland is firmly embedded in the control mentality that is so central to the relationship between scientific resource managers and nature. Wetland managers have attempted any number of assaults on purple loosestrife, ranging through pulling, mowing, disking, planting competitors (in one case, the non-native Japanese millet), chemical applications, and the introduction of European insect predators. The long-term ecological effects of many of these control methods are not known, and thus the act of cleansing nature may simply provide for the replacement of one form of "pollution" with another.[5] It is therefore the task of the wetland manager to create socially viable categories of desirability and undesirability, of acceptability and unacceptability and, in the ultimate God trick, a series of definitions for what counts as nature (i.e., the non-native purple loosestrife is out, but equally non-native beetles and Japanese millet are in). Only the disinterested discipline of science can make a claim to know nature objectively and, by extension, to act in the best interests of both nature and human society. As an arbiter between the social and the natural worlds, the purple loosestrife scientist not merely seeks to know nature but becomes a savior of the social order against all kinds of polluting influences.

The Problem with Exclusion: Nature as Commodity; Nature as Dirt

> *Man is in the habit of valuing things according to how well they serve his purposes. . . . Given his need for objects and his use for them, he draws the conclusion that they have been created to serve him. . . . Why should he not ignore a plant that is useless to him and dismiss it as a weed, since it really does not exist for him?*
>
> —GOETHE

Human society, both contemporary and ancient, is rife with conceptual boundaries and divisions. The distinction between the mind and the body, the separation of civilization from the wilderness, and the borderline between nature and culture have all been central to Western thought at least since the Renaissance. Such boundaries are, of course, more permeable than we like to admit: weeds and insects enter the domesticated garden, and wild animals dance in and out of the forest at the edge of the domesticated farm. Nevertheless, the conceptual boundaries remain fixed, at least in the human mind, and transgressions are not permitted to alter the sanctified categories of being.

Inevitably, the creation of boundaries and the placing of natural entities within these conceptual categories is entirely a human construct; they would not exist if we did not exist. As the anthropological work of Mary Douglas has shown us, the notion of boundary transgression, or pollution, has its origin in the social construction of reality. If culture "provides in advance some basic categories, a positive pattern in which ideas and values are tidily ordered" and "above all . . . has authority" (1966:38–39), then uncleanness, or pollution, "is matter out of place," and "we must approach it through order" (40). According to Douglas, "uncleanliness or dirt is that which must not be included if a pattern is to be maintained" (40).

While Douglas's earlier work concentrates on purity rituals and taboo in so-called primitive societies, she has applied her cultural theory of pollution to the secular and supposedly rational tenets of contemporary Western society. Pollution ideas are, according to Douglas and her colleague Aaron Wildavsky, "an instrument of control" that gives "the central establishment . . . the monopoly of explaining the natural order" (1982:47). Thus, "from the point of view of the central political establishment, the socially inferior are morally and physically contaminating, to be segregated and forcibly confined, punished if they try to break out" (47).[6]

While the relevance of socially constructed pollution categories to the governance of human relationships is obvious to any student of racism and ethnocentrism—particularly in the case of segregationist discourses—there remains the question of how explicit social categories tend to influence the concept of pollution in the natural world. One may fairly argue, for example, that the toxicity of DDT was not socially constructed and that its subsequent ban in North America was of immeasurable benefit to wildlife in general and raptors in particular. While I would not contest this point directly, one could suggest that the physical presence of DDT in the natural environment was/is a cultural response to another form of "invader": it belonged because the pollution of crop-eating insects did not. When a nature without birds and, more important, a civilization without humans was contemplated by Rachel Carson, only then was DDT removed and marginally safer alternatives found to control insects. The categories that make up what is natural and what is pollution are whatever humans want them to be and, quite often, whatever those in power assume they must be. As Douglas and Wildavsky remind us, "nature is what the center establishment sees as natural" (1982:47).[7]

With the latter point in mind, we must remember that the mere mention of the word "weeds" in the human context has evoked a whole range of metaphorical association with the battle to control nature. As the earlier biblical example illustrates, deep antipathies toward weeds run to the roots of our cultural origins as agriculturalists. Indeed, from the farmer's field to the suburban garden, the chemical industry and its patrons have spared no expense to rid nature of the first stages of plant colonization and succession. The war on weeds has—particularly in the case of the lawn manicure industry—extended far beyond the realm of necessity and, as Evernden reminds us, weeds have "become noxious not because of their physiological effect on cattle, but because of their conceptual effect on suburbanites; they are a *pollutant*. They are intrusions into the order of the lawn, and into the domain of human willing. Clearly then, "as 'natural' (wild) entities which must be excluded, weeds are dirt, as is the rest of nature" (1992:119).

But the war on purple loosestrife is apparently conducted on behalf of nature rather than aesthetic sensibility, an attempt to liberate the biotic community from the tyrannical influence of a life-destroying invasive weed. Its practitioners portray purple loosestrife control as an environmental initiative intended to save the wetland environment rather than control it. Accordingly, the purple loosestrife literature, scientific and otherwise, dutifully discusses the impacts of the weed on endangered species and threatened biodiversity more generally.[8] Purple loosestrife is a form of pollution according to the scientific community, and all of nature suffers under its pervasive influence.

Yet, in spite of the perceived and actual ecological effects of purple loosestrife, it is apparent that social definitions of pollution *have* influenced the scientific effort to "liberate" nature from purple loosestrife. In particular, the scientific management of purple loosestrife has remained true to its philosophical origin as an example of imperial ecology, an instrument used to dominate and control nature to the satisfaction of human desires. The ecologist Mark Anderson has pointed out, for example, that "birds, particularly game birds and waterfowl, provide the bulk of the justification for loosestrife management" (1995:227). However, no species other than the canvasback was identified in the Thompson paper as endangered in any way by purple loosestrife. Similarly, Thompson also discussed the impact of purple loosestrife on fur-bearing mammals at great length, though none of the species (muskrat, mink) can be considered threatened in North America. Purple loosestrife does, however, threaten the economics of exploiting such species, and Thompson carefully outlines the millions of dollars that will be lost to the economy of the Midwestern United States due to any loss of hunting, trapping, and recreation revenues due to a decline in the production of the wetland resource (Thompson, Stuckey, and Thompson 1987:43).

I do not mean to suggest here that human beings should never interfere with the natural world. In certain cases—particularly where endangered species and ecological communities are clearly at risk—a management intervention to control the impact of invasive species may be warranted. But, in the case of purple loosestrife, the crux of the matter is clearly not the preservation of a wetland community but the maintenance of a social and economic pattern that demands the dominance of human interests and influence in the natural landscape. Reptiles and amphibians, arguably the life forms that have suffered the most dramatic decline because of the commercial destruction of wetlands, have hardly been discussed in the purple loosestrife literature. Even the rhetoric of preserving the native plant community against an exotic invader rings hollow when one considers that wetland managers have struggled for fifty years to remove native stands of cattails to encourage abundant waterfowl populations (see Hager and McCoy 1998; Solberg and Higgins 1993), a process that is remarkably similar in concept and content to the duck-producing purple loosestrife campaign.[9] For the hunting groups that have been supporting purple loosestrife control and for their scientific allies in the universities and government agencies devoted to wildlife management, the wetland exists merely as a waterfowl-producing factory, and anything, be it purple loosestrife or cattails, that threatens this assigned social and economic role must be exterminated as a form of pollution regardless of the effect on the wider wetland community.[10] This obsession with managing wetlands "pollutants" to produce a maximum sustainable yield of a desired species/commodity reinforces the basic tenets and conceptual categories of a consumerist industrial society as "the preoccupation with productionism that has characterized so much parochial Western discourse and practice" becomes "hypertrophied into something quite marvelous: the whole world is remade into the image of commodity production" (Haraway 1992:297).[11]

In addition, the transgressive ontological boundaries erected by the extremely anthropocentric categories of North America's commodity culture prevent a more complete analysis of its relationship to the purple loosestrife "problem." Alfred Crosby has illustrated that exotic species have not historically been invaders in their own right but were instead followers in the wake of European expansion. According to Crosby, "the success of the portmanteau biota and of its dominant member, the European human, was a team effort by organisms that had evolved in conflict and cooperation for a long time" (1986:293). In other words, invasive organisms entered North American as a result of the expanding social, economic, and biological influence of European humans on the "new" continent. They are the by-products of a vicious colonial invasion that began over five centuries ago.

The more recent spread of purple loosestrife throughout North America (along the disturbed soil regimes associated with industrial infrastructure such as canals and superhighways) can be understood as a continuation of this invasion, the outgrowth of ongoing neo-European economic and social expansion in North America as opposed to the premeditated destruction of wetland plant communities by an external "vicious invader."[12] Furthermore, the extraordinary historical loss of wetlands throughout North America must also be attributed to the expanding engine of human enterprise rather than the introduction of purple loosestrife, though I have not seen any papers authored by weed scientists that discuss the spread of invasive commercial developments throughout the remaining wetlands of North America.

Given this new perspective, the purple loosestrife control effort must be seen in a new light: it acts not to save nature but to legitimize the commodity interests of the dominant culture by simultaneously mitigating its worst effects and by ensuring a continued bountiful harvest. The scientific manager's role in this world (i.e., civilization) assault on Earth (i.e., nature) is that of the "sane assassin," an emotionless defender of center values against the depredations of the uncontrollable wildness of nature.[13] The scientific manager is the emissary of the center who operates at the margin between nature and society, methodically providing the "cleanest" methods to kill insects, coyotes, wolves, weeds, and other varmints that pollute the social order of the productive farm, hunting area, town, or city. Purple loosestrife control is therefore not an act of preserving wetlands in the face of an alien invader. It is, rather, an assertion of power by human civilization over nature, and, as such, it reinforces the images of perfection that constitute the collective human construction of a socialized and, sadly, a sanitized natural world.

NOTES

1. An earlier version of this essay was published in the electronic Journal of Sociology 2 (September 1998), www.sociology.org. It also appeared in *Undercurrents: A Journal of Critical Environmental Studies* 9 (1997): 28–31.
2. Percent cover refers to the percentage of a given area, or a given stand of vegetation, covered by particular plant species. A genet is an individual plant that is composed of a number of nearly identical repeated units (i.e., one genet of purple loosestrife will tend to have many identical flowering stems).
3. See Keddy 1988; Hight and Drea 1991; Thompson 1991; DeClerck-Floate 1992; Becker and Welling 1993; Benckhuysen and Simser 1993; Blossey 1993; Haber, Keddy, and White 1993; Malecki et al. 1993; Manguin et al. 1993; Blossey and Schroeder 1995; Hight et al. 1995.
4. Lewontin provides many examples of these basic assumptions. Much like Worster, he argues that Darwin's economy of nature is very much borrowed from the political economy of late-nineteenth-century England. He also argues that the emerging political importance of individualism in the eighteenth century allowed an atomist worldview to emerge among European scientists, where the whole of any phenomenon can only be understood as the sum of its parts. This worldview also demands that scientists make clear distinctions between causes

and effects, even where these distinctions are not clear (see Lewontin 1991:9–16). Lewontin argues throughout the rest of the book against the emphasis in genetic science on the causal properties of individual genes; cancer is not, Lewontin suggests, caused by individual genes but is instead the product of interactions among genes, the organism, and the environment.

5. For example, scientific comments on the use of herbicides to control purple loosestrife reveals an astonishing "come what may" attitude to the release of synthetic chemicals. One body of research (Malecki and Rawinski 1985) cites several studies suggesting glyphosate has a low toxicity to fish, aquatic vertebrates, and birds but could affect young fish. The lack of a definition for "low toxicity" and the use of vague language suggest a lack of precise knowledge regarding the application of herbicides. Referring generally to herbicide use to control purple loosestrife, a prominent Canadian scientist suggests that "there remains an element of unknown risk through releasing these agents into the environment" (Keddy 1992:27). Skinner, Fuge, and Rendall (1994) further suggest that herbicides "provide only short-term control and have potential negative impacts on aquatic sites with repeated use" (20).

   To be fair, biological control using imported insects apparently does not present the same level of risks as chemicals because of its good record of success and low measure of ecological risk (Harris 1990; John Laing, personal correspondence, October 1995). As of 1980, 192 species of control agents had been established on 82 introduced plant species throughout the globe with only a few isolated attacks on native species (Hight and Drea 1991). There is, however, always an element of the unknown when dealing with complex living systems.

6. I am conscious of the fact that Douglas and Wildavsky use their analysis of pollution to question the environmental movement's construction of ecological collapse models. By using this material to support my arguments concerning purple loosestrife, I am not turning coat on the movement but suggesting the need to develop more compelling arguments than those that are purely technical in their orientation.

7. Nevertheless, the monopoly of central power structures on the definition of pollution categories may not be as rigid as Douglas and Wildavsky suggest. Fringe environmental groups opposing dam construction as a form of pollution in pristine river basins may oppose the powerful interests of government and business. Richard Grove's excellent work on the development of an early environmentalism among scientists in the colonial peripheries of the late European empires, far from the central influence of the intelligentsia in Paris and London and often in direct opposition to the commercial interests of the imperial center, suggests that too broad an association between science and power may be historically inaccurate (see Grove 1995, 1997). Thus, while I am sympathetic in a general sense to Douglas and Wildavsky's argument, I do not think it should be applied too broadly. I use it here only as a basic premise to discuss the particular case of purple loosestrife and the more general practice of scientific managerialism within natural resource bureaucracies.

8. Thompson, Stuckey, and Thompson (1987) discusses the bog turtle and the canvasback.

9. Indeed, without the "invasive species" rallying cry, the cattail managers are somewhat more forthright in their motives, one scientist declaring that "chemically created openings . . . are an acceptable management tool to create wetland openings that enhance waterfowl *use* and *production*" (abstract in Solberg and Higgins 1993, my emphasis).
10. A similar example came to the fore during the winter of 1995–1996, as the Ontario Federation of Anglers and Hunters ran a campaign to save Ontario's deer from the particularly harsh seasonal weather conditions. Donations were solicited so that feed could be distributed throughout the forests of the province, presumably so that enough deer would be available to be shot for recreation the following autumn. Again, it seems that saving the deer is not the issue so much as maintaining the productive supply. Paradoxically, in a fax addressed to a recent forum I attended on the ethics of deer culls to protect vegetation in provincial parks, OFAH suggested that southern Ontario was overpopulated with deer and that hunters should be allowed access to protected areas to rectify this pressing problem. Furthermore, it seems that the ducks that need to be saved from purple loosestrife are also threatened by the hunters whose representative organizations are supporting eradication campaigns. Hunters unable to identify waterfowl in the field may be shooting rare species and have even been known to shoot hawks, herons, grebes, and shorebirds. See McKay 1996.
11. Andrew Light and Eric S. Higgs have written a fascinating paper on the relationship between restoration projects and capitalist commodification. Their particular focus is the corporatization of restored images of nature to act as an apology for continued exploitation of nature and as a pristine image to sell products. See Higgs and Light 1996.
12. See Thompson, Stuckey, and Thompson 1987 for an early expansion history along canals, and Wilcox 1994 for an account of superhighways as a major agent of spread into the western United States.
13. I am borrowing here from Dennis Lee the concept of a continual struggle between world (civilization) and Earth (nature). Lee discusses Michael Ondaatje's poetic representation of this battle in the person of the law enforcement officer Pat Garrett, a "sane assassin" who controls the wild unpredictability of the outlaw Billy the Kid (1977:37–44). I am suggesting that the scientist is modern society's ontological law enforcement agent.

REFERENCES

Anderson, Mark G. 1995. Interactions between *Lythrum salicaria* and native organisms: A critical review. *Environmental Management* 19 (2): 225–231.

Batra, S. W., T. D. Schroeder, P. E. Boldt, and W. Mendl. 1986. Insects associated with purple loosestrife (*Lythrum salicaria*) in Europe. *Proceedings of the Entomological Society of Washington* 88: 748–759.

Becker, Roger L., and Charles H. Welling. 1993. Reduction of purple loosestrife establishment in Minnesota wetlands. *Wildlife Society Bulletin* 21 (1): 56–64.

Benckhuysen, J. A., and W. L. Simser. 1993. Efficacy of tryclopyr amine in controlling purple loosestrife. *Garden's Bulletin* 12 (2): 23–38.

Blossey, B. 1993. Herbivory below ground and biological weed control: Life history of a root-boring weevil on purple loosestrife. *Oecologia* 94: 380–387.

Blossey, B., Stephen D. Hight, Richard A. Malecki, and Dieter Schroeder. 1994. Host specificity and environmental impact of the weevil *Hylobius transversovittatus*, a biological control agent of purple loosestrife. *Weed Science* 42: 128–133.

Blossey, B., and Dieter Schroeder. 1995. "Host specificity of three potential biological weed control agents attacking flowers and seeds of *Lythrum salicaria. Biological Control.* 5: 47–53.

Crosby, Alfred. 1986. *Ecological Imperialism: The Biological Expansion of Europe, 900–1900.* Cambridge: Cambridge University Press.

DeClerck-Floate, Rosemarie. 1992. *The Desirability of Using Biocontrol Against Purple Loosestrife in Canada.* Lethbridge: Agriculture Canada.

Delfosse, E. S. 1984. *Proceedings of the VI International Symposium on Biological Control of Weeds.* Ottawa: Agriculture Canada.

Douglas, Mary. 1966. *Purity and Danger: An Analysis of the Concepts of Pollution and Taboo.* London: Routledge.

Douglas, Mary and Wildavsky, Aaron. 1982. *Risk and Culture: An Essay on the Selection of Technical and Environmental Dangers.* Berkeley: University of California Press.

Ducks Unlimited. n.d. *Purple Loosestrife.* An information pamphlet published by Ducks Unlimited Canada, with the support of the Canadian Wildlife Service and the Canadian Wildlife Federation.

Evernden, Neil. 1992. *The Social Creation of Nature.* Baltimore: Johns Hopkins University Press.

Federation of Ontario Naturalists. 1992. *A Report on the 1992 Purple Loosestrife Survey.* Don Mills: FON.

Gleik, James. 1987. *Chaos: The Making of a New Science.* New York: Penguin.

Goethe, Johanne Wolfgang von. 1988. *Scientific Studies.* Ed. and trans. Douglas Miller. New York: Suhrkamp.

Grove, Richard. 1995. *Green Imperialism: Colonial Expansion, Tropical Island Edens and the Origins of Environmentalism, 1600–1860.* New York: Cambridge University Press.

——. 1997. *Ecology, Climate and Empire: Colonialism and Global Environmental History, 1400–1940.* Cambridge: White Horse.

Haber, Eric, Cathy Keddy, and David J. White. 1993. *Invasive Plants of Natural Habitats in Canada. Canadian Wildlife Service.* Ottawa: Environment Canada.

Hager, Heather A., and Karen D. McCoy. 1998. The implications of accepting untested hypotheses: A review of the effects of purple loosestrife (*Lythrum salicaria*) in North America. *Biodiversity and Conservation* 7 (8): 1069–1079.

Haraway, Donna. 1988. Primatology is politics by other means. In *Feminist Approaches to Science.* Ed. Ruth Bleier. New York: Pergamon.

———. 1992. The promises of monsters: A regenerative politics for inappropriate/d others. In *Cultural Studies.* Eds. Laurence Grossberg, Cary Nelson, and Paula Treichler. New York: Routledge.

Harris, Peter. 1990. Environmental impact of introduced biological control agents. In *Critical Issues in Biological Control,* 289–300. Eds. L. Ehler, M. Mackauer, and J. Roland. Andover: Intercept.

Hays, Samuel. 1959. *Conservation and the Gospel of Efficiency: The Progressive Conservation Movement, 1890–1920.* New York: Atheneum.

Higgs, Eric S., and Andrew Light. 1996. The politics of ecological restoration. *Environmental Ethics* 18: 227–247.

Hight, Stephen D., Bernd Blossey, John Laing, and Rosemarie DeClerck-Floate. 1995. Establishment of insect biological control agents from Europe against *Lythrum salicaria* in North America. *Environmental Entomology* 24 (4): 967–977.

Hight, Stephen D., and John Drea. 1991. Prospects for a classical biological control project against purple loosestrife. *Natural Areas Journal* 11 (13): 151–157.

Keddy, Cathy 1988. The urgency for purple loosestrife management in Ontario. In *Wetlands: Inertia or Momentum.* Eds. Michael Bordecki and Nancy Patterson. Don Mills: Federation of Ontario Naturalists.

———. 1992. *A Universal Manual for Purple Loosestrife Control.* Peterborough: Ontario Federation of Anglers and Hunters.

Kuhn, Thomas. 1962. *The Structure of Scientific Revolutions.* Chicago: University of Chicago Press.

Lee, Dennis. 1977. *Savage Fields: An Essay in Literature and Cosmology.* Toronto: Anansi.

Lewontin, Richard. 1991. *The Doctrine of DNA: Biology as Ideology.* London: Penguin.

McKay, Barry Trent. 1996. Bag limits a joke if hunters can't identify waterfowl. *Toronto Star.* October 20, E4

Malecki, Richard A., Bernd Blossey, Stephen D. Hight, Dieter Schroeder, Loke T. Kok, and Jack R. Coulson. 1993. Biological control of purple loosestrife. *Bioscience* 43 (10): 680–684.

Malecki, R. A., and T. J. Rawinski. 1985. New methods for controlling purple loosestrife. *New York Fish and Game Journal* 32 (1): 81–87.

Manguin, Sylvie, Richard White, Bernd Blossey, and Stephen D. Hight. 1993. Genetics, taxonomy, and ecology of certain species of *Galerucella. Annals of the Entomological Society of America* 86 (4): 398–410.

Skinner, Luke C., Ellen L. Fuge, and William J. Rendall. 1994. *Minnesota's Purple Loosestrife Program: History, Findings, and Management Recommendations.* St. Paul: Minnesota Department of Natural Resources.

Solberg, K. L., and K. F. Higgins. 1993. Effects of glyphosate herbicide on cattails, invertebrates, and waterfowl in South Dakota wetlands. *Wildlife Society Bulletin* 21 (3): 299–307.

Thompson, Daniel Q. 1991. History of purple loosestrife biological control efforts. *Natural Areas Journal* 11 (3): 148–150.

Thompson, D. Q., R. L. Stuckey, and E. B. Thompson. 1987. *Spread, Impact and Control of Purple Loosestrife in North American Wetlands.* U.S. Fish and Wildlife Research 2. U.S. Fish and Wildlife Service.

Westman, Walter. 1990. Park management of exotic species: problems and issues. *Conservation Biology* 4 (3): 251–260.

Wilcox, Douglas A. 1994. Migration and control of purple loosestrife (*Lythrum salicaria*) along highway corridors. *Environmental Management* 13 (3): 365–370.

Worster, Donald. 1994. *Nature's Economy: A History of Ecological Ideas.* Cambridge: Cambridge University Press.

Zajonc, Arthur. 1993. *Catching the Light: The Entwined History of Light and Mind.* New York: Oxford University Press.

NINE

Restoration, Autonomy, and Domination

ANDREW LIGHT

Introduction

ECOLOGICAL RESTORATION is the practice of re-creating ecosystems that have been previously destroyed, largely because of anthropogenic causes. Such endeavors range from small-scale park projects, such as those currently under way in the Chicago forest preserves and Prospect Park in Brooklyn, New York, to the huge multibillion-dollar restoration of the Florida Everglades that has been under way since the Clinton administration. Many nonanthropocentrists in environmental ethics who have written about ecological restoration have been skeptical of the propriety of this practice, if not highly critical of it. Why? The initial reasons are essentially embedded in an ontological claim that restored environments are not natural and hence can never equal in value the natural entities they are supposed to replace. The argument goes something like this: If something is the product of an act of human creation (as a restored landscape is), then it has an irreducible anthropogenic component. As such, a restored environment can never count as the sort of thing that contains natural value, nonanthropocentrically conceived, since one reason that nature may have such a value is its independence from human acts of creation. From the perspective of a form of nonanthropocentric holism (the theory that obligations to the nonhuman natural world are to whole ecosystems and not to individual entities), humans and naturally evolved species and ecosystems are the only things to which we owe some degree of direct moral consideration, and therefore a restored environment is at best an artifact that can never substitute for the value that such a view would attribute to original nature.

From this perspective, restorations are not necessarily bad things. They are simply second best, if what one values most is some form of "authentic" nature (whatever one believes that actually means). It is the dependence of restorations on humans that gives them the status of mere artifacts, as opposed to the independence of nature, which in part makes it a candidate for direct moral consideration. At worst, however, restorations can represent, in this view, the tyranny of humans over nature, especially if one sees the independence of nature to be part of a justification for its autonomy. Eric Katz, who claims that nature is a subject akin to a human subject (as is once again apparent in his contribution to this volume) and hence the sort of thing whose autonomy should be respected, has articulated this sentiment most emphatically in arguing that, "the practice of ecological restoration *can only* represent a misguided faith in the hegemony and infallibility of the human power to control the natural world."[1] When we choose to engage in an ecological restoration, even for benevolent reasons, we are simply imposing our shortsighted human will on nature, in much the same way we do when we destroy an original ecosystem and develop something else on top of it. While our motivations may appear to be better in the case of, say, restoring a tall-grass prairie in a rural area of Ohio, we are, following Katz's line of reasoning, once again only mucking with the environment in an abusive way. While it is more akin to a benignly inspired bit of paternalism in a comparable human case, at the end of the day it is just another instance of unethical domination.

I have long disagreed with claims like this. My previous answer to such positions has been simply to set aside these kinds of arguments and focus on other issues involving ecological restoration that help to demonstrate its social and political importance.[2] But I do not think it wise to ignore such views. In addition to the philosophical arguments that must be mustered to respond to them, we should also consider whether the kind of environmental philosophy that produces such claims really provides the best context within which to assess practices such as ecological restoration, a practice that any charitable assessment of the bulk of projects under way today will surely reveal does more good than harm for the environment. I for one believe that environmental philosophers should be interested in more than simply describing the nonanthropocentric value of nature and be wary of any form of value theory that pushes a thesis concerning the supposed autonomy of nature so far that it winds up with counterintuitive environmental conclusions like the ones embraced by Katz.

What alternatives exist? One answer would be simply to abandon nonanthropocentric holism in environmental ethics altogether and thus avoid the questions of the potential autonomy of nature.[3] After all, the extreme position Katz takes up is only a live option with the fairly narrow theoretical constraints that have been produced within this branch of environmental philosophy. If moral consideration need only be extended to human subjects, then these are the only kinds of beings that potentially possess a quality like autonomy that should be respected in a moral sense. As such, the relative independence or dependence of ecological restorations from humans would be a moot point in their overall assessment. While this is the view I am generally inclined to take (for reasons altogether independent from the question of the autonomy of nature and its effects on our assessment of restoration), I will not pursue that line of argumentation here. Instead, I will argue that the position Katz lays out does not justify his critique of restoration on its own terms and that in turn, from a view that is agnostic with respect to the question of the autonomy of nature, we have ample resources to endorse the goals and practice of most ecological restorations. Restoration makes sense because, on the whole, it results in many advantages over mere preservation of ecosystems that have been substantially damaged by humans. As the last section of this chapter will show, some of those advantages have to do with the benefits of restoration for improving the human relationship with naturally evolved systems and entities and not just benefits to nature itself. But, first, how can we answer the sorts of criticisms offered of restoration by nonanthropocentrists like Katz?

Katz Against Restoration

As Katz describes it, there are actually two separable questions to put to advocates of restoration ecology: (1) Do we have an obligation to try to restore damaged nature? and (2) Do we have the ability to restore damaged nature? Katz argues quite forcefully that we do not have the ability to restore nature because what we actually create in ecological restorations are humanly produced artifacts and not nature. Based on this claim, he assumes that the first question—whether we have an obligation to try to restore nature—is moot. Katz's logic is simple: we do not have an obligation to do what we can't in principle do.

But even if we were to grant Katz the argument that it is impossible to restore "nature," in the sense that Katz understands what is meant by this term, it might still be the case that we have moral obligations to *try* to restore nature. How could this be true? There are a number of reasons, which I will raise below, but for now consider that what we are really restoring with restoration is not necessarily nature itself but some kind of relationship with nature (whether actually there in the restoration, adjacent to it, or as a more abstract idea). But before fully explicating this position, we need first to better understand Katz's arguments.

There are five separable, but often overlapping, claims in Katz's work on restoration opposing both the suggestion that we can restore nature and the practice of trying to restore nature. They are listed below in the order that they arise in Katz's work, each accompanied with an example of supporting evidence from Katz's various papers on restoration.

The Duplicity Claim (KR1): "I am outraged by the idea that a technologically created 'nature' will be passed off as reality."[4]

The Arrogance (or "Hubris") Claim (KR2): "The human presumption that we are capable of this technological fix demonstrates (once again) the arrogance with which humanity surveys the natural world."[5]

The Artifact Claim (KR3): "The re-created natural environment that is the end result of a restoration project is nothing more than an artifact created for human use."[6]

The Domination Claim (KR4): "The attempt to redesign, re-create, and restore natural areas and objects is a radical intervention in natural processes. Although there is an obvious spectrum of possible restoration[s] . . . all of these projects involve the manipulation and domination of natural areas. All of these projects involve the creation of artifactual realities, the imposition of anthropocentric interests on the processes and objects of value. Nature is not permitted to be free, to pursue its own independent course of development."[7]

The Replacement Claim (KR5): "If a restored environment is an adequate replacement for the previously existing natural environment [which of course for Katz it can never be], then humans can use, degrade, destroy, and replace natural entities and habitats with no moral consequence whatsoever. The value in the original natural entity does not require preservation."[8]

Here, I will focus on KR4, the domination claim, which is perhaps the one that comes up the most throughout all of Katz's restoration papers. I will argue that the rest of Katz's claims can be conceded as long as KR4 can be independently answered.

KR 1–3 and 5 can be ignored in rejecting Katz's position so long as we are prepared to concede for now one important supposition backing up all his claims. This is Katz's ontological assumption that humans and the rest of nature can be meaningfully separated, thus grounding the argument that restored nature is an artifact, a part of human culture, rather than a part of nature. As Katz has admitted, he is a nature-culture dualist.[9] If one rejects this overall ontological and metaphysical view about the separation of humans from nature, then one may reject most of Katz's objections to restoration.

But I will accept here, even though I disagree with it, Katz's underlying assumption that restored environments do not reproduce nature, that is to say, it does not reproduce whatever value Katz wishes to attribute to nature. But even if I grant the point that restored nature is not really nature, KR4 is still false because it is arguably the case that restoration does not dominate nature in any coherent sense but often instead helps nature, endowed now with some sense of autonomy and a normative requirement that we respect that autonomy, to be free of just the sort of domination by humans that Katz is worried about. The reasoning here is straightforward enough. If I can show that restorations are valuable for nature, even if I concede that they do not re-create nature, then the various motivations for restoration will distinguish whether a restoration is duplicitous (KR1) or arrogant (KR2). If one restricts the discussion to what I have called elsewhere "benevolent restorations" (those intended to rectify a past harm done to nature) rather than a more malevolent form of restoration (those intended to justify destruction of nature on the assumption that a restoration can always bring back the full natural value of a destroyed area), then neither KR1 or KR2 would be realized because, in principle, such restorations are not trying to fool anyone and are not necessarily arrogant.[10]

Further, and more simply, conceding Katz's ontological claim about the distinction between nature and culture eliminates the significance of KR3—since we no longer care that what is created may or may not be an artifact—as well as the possibility of KR5 being realized, since we have given up hope that a restoration could ever actually serve as a replacement for "real" nature. (Notice, too, that KR5 would only apply if one wishes to defend malicious restorations, which I doubt any environmentalist is willing to do.)

Now, back to the domination claim. KR4 is a claim that could hold even for a view that conceded Katz's nature-culture distinction. The reason for Katz would be that even a failed attempt to duplicate natural value—or create something akin to nature, while conceding that in principle humans can never restore "real" nature—could still count as an instance of domination as Katz has described it. An *attempt* at restoration, in Katz's logic, would still prohibit nature from ever being able to evolve in whatever way it would have evolved without some form of human interference. The reason is that, for Katz, restoration is always a substitute for whatever would have occurred at a particular site absent human interference. The idea is that even if humans can produce a valuable landscape of some sort on a denuded acreage, this act of production is still an instance of domination over the alternative of a natural evolution of this same acreage, even if a significant natural change would take ten times as long as the human induced change and would be arguably less valuable for the species making use of it. Restoration ecology, in this scheme, represents a limitation on the autonomy of nature. Still, one can muster a number of arguments against KR4 (I will provide four) and still play largely within Katz's biggest and most contentious assumption about the ontological status of restored nature.

1. We can imagine cases where nature cannot pursue its own interests (however one wishes to understand this sense of nature having interests) because of something we have done to it that must be rectified by restoration. For example, many instances of restoration are limited to bioactivation of soil that has become contaminated by one form or another of hazardous industrial waste. If restoration necessarily prohibits nature from being "free," as KR4 maintains, then how do we reconcile the relative freedom that bioactivation makes possible with this claim? Restoration need not determine exactly what grows in a certain place but may in fact simply be the act of allowing nature to pursue its own interests again rather than shackling it to perpetual human-induced trauma. In many cases of restoration, this point can be driven home further when we see how anthropogenically damaged land (or soil) can be uniquely put at risk of invasion by anthropogenically introduced exotic plants.

The South African ice plant, an exotic in southern California that destroys the soil it is introduced to, is highly opportunistic and can easily spread onto degraded land, thus ensuring that native plants will not be able to reestablish themselves. I highlight here this contentious native-exotic distinction because I suspect that, given Katz's strong nature-culture distinction, he would necessarily have to prefer a landscape of native plants to a landscape of exotics where the existence of the exotics is the result of an act of human (cultural) interference in nature. If the original nature at such a site were never put at risk of invasion by exotics introduced by humans, then we can be relatively sure that those exotics would not have made inroads onto the site. Allowing nature to pursue its own interests, given prior anthropogenic interference, thus involves at least as strong a claim to protect it from further anthropogenic risk through restoration practices as the case Katz makes for leaving it alone.

2. Going back to a point made earlier, even if we do agree with Katz that restorations only produce artifacts, can't it still be the case that the harm we cause to nature requires us to engage in what we would have to term, following Katz, "attempted restorations"? It simply does not follow from the premise that something is more natural when it is relatively free of human interference that we should conclude that therefore we must always avoid interfering with nature. It is a classic premise of nonanthropocentric holism that some interference is warranted when we are the cause of an imbalance in nature: for example, hunting white-tailed deer is thought to be permissible since humans have caused that species' population explosion. If such interventions as hunting are permissible as an aid to help to rectify the balance of

nature, then why are there not comparable cases with the use of restoration as an aid for the "original," "real" nature? We can even imagine that such cases would be less controversial than holist defenses of hunting.

It turns out in fact that there are good cases where restoration, even if it results in the production of an artifact, does not lead to the domination or infringement of autonomy described by Katz. Imagine the case where a restoration project will restore a corridor between two wilderness preserves. If there is positive natural value in the two preserves that is threatened because wildlife is not allowed to move freely between them, then restoration projects that would restore a corridor (such as removing roads, for example) would actually not only be morally permissible but possibly ethically required depending on one's views of the value of the nature in the preserves. This is not restoration as a second-best to preservation or a distraction from preservation, it is restoration as an integral and critical part of the maintenance of natural value. So, even if we agree with Katz that humans cannot really restore nature, it does not follow that they ought not to engage in restoration projects that actually repair the damage caused by past domination rather than furthering that domination.

Given objections like the two discussed so far, it is important to try to get a better handle on exactly what sort of damage is caused by domination in the sense described by Katz. It turns out that the worst damage to nature for Katz is domination that prevents the "self-realization" of nature. "The fundamental error is thus domination, the denial of freedom and autonomy. Anthropocentrism, the major concern of most environmental philosophers, is only one species of the more basic attack on the preeminent value of self-realization. From within the perspective of anthropocentrism, humanity believes it is justified in dominating and molding the nonhuman world to its own human purposes."[11] Thus the problem with restoration is that it restricts natural self-realization by forcing nature onto a path that we would find more appealing.

3. With this clarification, we can then further object to Katz that his sense of restoration confuses the practice with mitigation, for example, the practice of creating new wetlands where none had existed before in order to make up for the loss of an original wetland elsewhere. The force of the charge of domination is that we mold nature to fit our "own human purposes." But with restoration as a practice, the point of most scientific disputes over it is precisely that anything does not go. While there is always some variability in what can be restored at a particular site (what period, after all, do we restore to?), we cannot restore a landscape just any way we wish and still have a good restoration in scientific terms. We are also bound in the context of restoration, as was mentioned before, of restoring to some preexisting state even if we are unsure which particular historical state we ought to restore to. If that

is the case, then the broadly construed historical and scientific boundaries of restoration limit the purposes to which we can put a restoration.

If Katz objects that, when we restore a denuded bit of land, we are at least making something that fits our need of having more attractive "natural" surroundings—an argument that Katz often makes—we can reply that, because of the constraints placed on restoration as opposed to mitigation, the fact that we find a restored landscape appealing is only contingently true. It can clearly be the case (and there are plenty of empirical examples) that what we must restore to is not the preferred landscape of most people. The Chicago Wilderness project (a joint endeavor of over forty public and private agencies in Chicago to restore the forest preserves around the city, successfully restoring over fourteen thousand acres so far) was in fact in some trouble over exactly these kinds of worries. Many people see this project as involving the destruction of aesthetically pleasing forests in order to restore the original oak savannas (the kind of landscape the project's organizers claim was in place before white settlement). But because a restored landscape can never necessarily be tied only to our own desires (since our desires are not historically and scientifically determined in the same way as the parameters of a restoration), those desires cannot actually be the direct cause of any restriction on the self-realization of nature.

4. Finally, we must wonder at this value of self-realization and the assumption of the nature of the autonomy of nature that underlies this view. Setting aside the inherent philosophical problems with understanding what this claim to self-realization means in the case of nature, one has to wonder how we could know what natural self-realization would be in any particular case and why we would totally divorce a human role in helping to make it happen if we could discern it. In an analogous case, involving two humans, we do not say that a human right to (or value of) self-realization is abrogated when a criminal who harms someone is forced to pay restitution. Even if the restitution is forced against the will of the victim, and even if the compensation in principle can never make up for the harm done, we would not say that somehow the victim's self-realization has been restricted by the act of restitution by the criminal. Again, there seems to be no clear argument here for why the moral obligation to try to restore has been diminished by Katz's arguments that we do not have the ability to restore "real" nature or that we are passing off an artifact as nature.

Restoring Environmental Philosophy

If I am justified in setting aside the rest of Katz's claims (KR 1–3 and 5) by accepting his claim that humans really cannot restore "real" nature, and if my objections to KR4 hold true, then what sort of conclusions could we draw about the role of philosophy in sorting out the normative issues involved in restoration? As it turns out, Katz gives us an insight into figuring out the next step.

After explaining the harm we do to nature in the domination we visit upon it through acts of restoration, Katz briefly assesses the harm that we do to ourselves through such actions: "A policy of domination transcends the anthropocentric subversion of natural processes. A policy of domination subverts both nature and human existence; it denies both the cultural and natural realization of individual good, human and nonhuman. Liberation from all forms of domination is thus the chief goal of any ethical or political system."[12]

Though not very clearly explained by Katz, this intuition represents a crucial point for proceeding further. In addition to connecting environmental philosophy to larger projects of social liberation, Katz here opens the door to a consideration of the consequences of restoration for humans and human communities. As such, Katz allows for the possibility that there is a value involved in restoration other than the value of the objects that are produced by a restoration. The problem with drawing this conclusion, though, is that this passage is also perhaps the most cryptic in all of Katz's work on restoration. What does Katz mean by this claim? How exactly does restoration deny the realization of an individual human, or cultural, good? This claim can only be made understandable by assuming that there is some kind of cultural value connected to nature that is risked through the act of domination or by otherwise causing harm to nature. But what is this value?

I think that the value Katz is alluding to here, though he never explores it seriously, somehow describes the value of that part of human culture that is connected to external, nonhuman nature. This is not simply a suggestion that we humans are part of nature (for the implication here is that we cause a disvalue to ourselves and to nature by acting *on* nature in some way, thus confirming Katz's nature-culture distinction) but rather that we have a *relationship with* nature that exists on a moral as well as physical terrain in such a way that our actions toward nature can reciprocally harm us. If this is the view implicit in this claim, then it is still consistent with much of the rest of Katz's larger views about the value of nature. We have a relationship with nature even if we are separable from it. Without fully explicating the content of that relationship, it seems that Katz is right in assuming that somehow the way in which we act toward nature morally implicates us in a particular way. In the same sense, when we morally mistreat other human beings, we not only harm them but harm ourselves (by diminishing our character, by implicating ourselves in evil, however you want to put it).

Now, if this assumption is correct, and if there is anything to the arguments I have put forward so far that there can be some kind of positive value to our interaction with nature, then doing right by nature from a nonanthropocentric perspective will have the reciprocal effect of morally implicating us in a positive value as occurs when we do right by other persons. From an anthropocentric perspective, it may simply be seen as a virtuous thing to do given that the way we treat the environment around us has effects on other people. Perhaps Katz would agree. Where Katz would disagree is with the suggestion I would add to this that some part of many kinds of restoration (if not most kinds) contains positive value. Aside from the other suggestions I have already made concerning the possible positive content of restoration, one can also consider that the relationship with nature that is implied in Katz's view has a moral content in itself that is not reducible to the value of fulfilling this relationship's concomitant obligations. The relationship between humans and nature imbues restoration with a positive value even if it cannot replicate natural value in its products. But understanding this point will require some explanation.

If I have a reciprocal relationship with another human (where I do right by this person and he or she does right by me), then, to generalize Katz's account, there is a moral content to both of our actions that implicates each of us as persons. Each of us is a better person morally because of the way we interact with each other in the relationship. But the relationship itself, or rather just the fact of the existence of the relationship, also has a moral content of its own (or what we would call a "normative content," meaning that the relationship can be assessed as being in a better or worse state) that is independent of the fulfillment of any obligations. If this point of the possible separation between the value of a relationship and the value of the fulfillment of obligations does not follow intuitively, imagine the case where two people act according to duty toward each other without building a substantive normative relationship between them. Consider the following example.

I have a brother to whom I am not terribly close. While I always act according to duty to him—I never knowingly do harm to him and I even extend special family obligations to him—I do not have a substantive relationship with him, that in itself has a normative content. Thus, if I do not speak to him for a year, nothing is lost (indeed, neither of us feels a loss) because there is no relationship there to maintain or that requires maintenance for normative reasons. But, if my brother needed a kidney transplant, I would give him my kidney unhesitatingly out of a sense of obligation—something I would not feel obliged to do for nonfamily members—even though I still do not feel intimately comfortable around him in the same way that I do with my closest friends. Our relationship as persons—that sense of intimate affection and care for another person that I have experienced with other people—has no positive value for me. (It isn't necessarily a disvalue; it is only a sense of indifference, a lack of closeness.) So, I can have interaction with another person, even interaction that involves substantial components of obligation and duty (and, in Katz's terms, I will never put myself in a position to dominate that other person), but still not have a relationship with him or her that involves any kind of positive value or has normative standards of maintenance.

I do not think that I have any obligation to have a relationship in this sense with my brother. In fact, I do not, even though my mother would like it if I did. But, if I did have a relationship with my brother in this sense, then it would have a value above and beyond the moral interaction that I have with him now (the obligations that I have to him that can be iterated) that would aid in a determination of our moral character.[13] If we had a relationship with normative content, a positive or negative value could be assessed if I lost touch with my brother or ceased to care about his welfare. (I could very well claim that it would be better for me to have such a relationship with him, but this would require another argument.)

Consider further that if I wanted to rectify or create anew a substantive normative relationship with my brother, like the relationship I have with several close friends, how would I do it? One thing that I could do would be to engage in activities with him, the same sorts of activities that I pursue with my friends now. I might work with him to put up a fence or help him plant his garden. I might begin to talk over my personal and professional problems with him. I might go on a long journey with him that demanded some kind of mutual reliance, such as white-water rafting or visiting a foreign city where neither of us spoke the native language. In short, though there are of course no guarantees, I could begin to have some kind of material relationship with him as a prelude to having some kind of substantive normative relationship with him. Many factors might mitigate the success of such a project, for one thing, the distance between us: he lives in our hometown of Atlanta and I live in New York City. So, if I were really serious about this project of building a relationship between us that had value independent of the value of the fulfillment of our mutual obligations to each other, which already exists, I'd have to come up with ways to bridge these interfering factors. Importantly, though, I couldn't form a substantive normative relationship with him merely by respecting his right of self-realization and autonomy as a person; I would have somehow to become actively involved with him.

Now, when we compare this case of estranged brothers to that of nature, many parallels arise. We know that we can fulfill obligations to nature in terms of respecting its autonomy and self-realization as a subject (in Katz's terms) without ever forming a substantive normative relationship with it. Assuming also that there is a kind of relationship with nature possible on Katz's scheme (for this is in part what we harm when we dominate nature), it is fair to say that a relationship consisting in positive value with nature is compatible with Katz's overall view of the human-nature relationship. Because he says so little about what our positive relationship to nature could be, he is in no position to restrict it a priori. We also know that, as in the case of the estranged brother, we need some kind of material bridge to create a relationship with nature in order to see that relationship come about.

How to build that bridge? Suggesting ways to overcome the gap between humans and nature (without necessarily disvaluing it) seems in part to be the restored role of an environmental philosophy in questions of ecological restoration. Certainly, as in the case of my brother, distance is a problem. Numerous environmental professionals have emphasized the importance of being in nature in order to care for it. Also, acts of preservation are important in order for there to be nature with which to have a relationship. But what about restoration? Can restoration help engender such a positive normative relationship with nature? It seems clear to me that it can. When we engage in acts of benevolent restoration, we are bound *by* nature in the same sense that we are obligated to respect what it once was attempting to realize before we interfered with it. In Katz's terms, we are attempting to respect it as an autonomous subject. But we are also bound *to* nature in the act of restoring. In addition to the ample sociological and anthropological evidence on the positive value with nature that is engendered in benevolent restoration, we can say that restoration restores the human connection to nature by restoring that part of culture that has historically contained a connection to nature. While it would take further argument to prove, I believe that this is the kind of relationship that is a necessary condition for encouraging people to choose to preserve the natural system and landscapes around them rather than trade them off for short-term monetary gains garnered by development. If I am in a normative relationship with the land around me (whether it is "real" nature or not), I am less likely to allow it to be harmed further.[14]

We can even look to Katz for help in completing this pragmatic task. We don't want restorations that try to pass themselves off as the real thing when they are really "fakes" (KR1) or are pursued through arrogance (KR2); neither are we interested in those that are offered as justifications for replacing or destroying nature (KR5). We would not want our human relationships to exhibit those properties, either. But even given the legacy of inhuman treatment of each other, we know that it is possible to restore human relationships that do not resemble the conditions described by the claims proposed in KR1, KR2, and KR5. There is, however, one possible worry to attend to concerning KR3: the artifact claim. While earlier I had said that the importance of KR3 is diminished by granting Katz's nature-culture distinction, there is a way that it can still cause us problems in grounding attempts at restoration in the positive value of strengthening the human-nature relationship.

Katz may object to my relationship argument by saying that, if we allow his claim that what has been restored is not really nature, then what we are restoring is not a cultural relationship with nature but, in a sense, only extending the artifactual material culture of humans. At best, all we can have with restoration is a relationship with artifacts, not nature. Maybe he will allow that we improve relations with each other through cooperative acts of restoration, but this is not the same as the restoration of a relationship with nature itself.

But it should be clear by now that Katz would be mistaken to make such an objection for several reasons stemming in part from my earlier remarks:.

1. Even if we admit that restored nature is an artifact and not real nature, restored nature can also serve as a conduit for real nature to free itself from the shackles we have previously placed on it. Restoration can allow nature to engage in its own autonomous restitution. Of the different sorts of restoration projects that I have sketched above, many amount to aids to nature rather than creations of new nature.

2. Even if restoration is the production of an artifact, these artifacts do bear a striking resemblance to the real thing. This is not to say that restorations can be good enough to fool us (KR1). Rather, it is simply to point out that an opportunity to interact with the flora and fauna of the sort most common in benevolent restorations will have the effect of increasing the bonds of care that people will have with nonrestored nature. If a denuded and abandoned lot in the middle of an inner-city ghetto is restored by local residents who have never been outside of their city, then it will give them a better appreciation of the fragility and complexity of the natural processes of nature itself should they encounter them. The fact that restorationists are engaged in technological processes does not necessarily mean that their practices do not serve the broader purpose of restoring a relationship with nature. Just as starting some form of mediated communication with my brother (such as email or regular phone calls) does not restore a fully healthy communicative relationship with him that could be found through face-to-face conversation, it still helps me to get used to the idea of some form of immediate and substantive communication.

3. And finally, if Katz persists in his worry that the act of restoration reifies domination by reaffirming our power over nature through the creation of artifacts, we can say that exactly the opposite is likely the case (at least in the case of benevolent restorations) where the goal is restoring the culture of nature if not nature itself. Restorationists get firsthand (rather than anecdotal

and textbook) exposure to the actual consequences of human domination of nature. A deeper understanding of the problems of bioactivating soil, for example, gives us a better idea of the complexity of the harm we have caused to natural processes. Becoming aware of that harm can be empowering in a much healthier way than Katz seems willing to admit in that it can empower us to know more precisely why we should object to the kinds of activities that can cause that harm to nature in the first place.[15]

It seems clear that benevolent restorations are valuable because they help us to restore our relationship with nature by restoring some kind of cultural relationship with the nonhuman natural world. This is true even if Katz is correct that restored nature has the ontological property of an artifact. Restoration is an obligation exercised in the interests of forming a positive community with natural systems, whether they possess the quality of autonomy or not, and thus is well within the boundaries of a constructive environmental philosophy.[16]

NOTES

1. Eric Katz, "The Problem of Ecological Restoration," *Environmental Ethics* 18 (1996): 222, my emphasis. See also "The Call of the Wild: The Struggle Against Domination and the Technological Fix of Nature," "The Big Lie: Human Restoration of Nature," "Artifacts and Functions: A Note on the Value of Nature," and "Imperialism and Environmentalism." All these papers are collected in Eric Katz, *Nature as Subject: Human Obligation and Natural Community* (Lanham, Md.: Rowman and Littlefield, 1997). I have drawn on the versions of these papers as they appear in that volume for this chapter. For a more updated defense of his views on restoration ecology since the publication of the original version of this chapter in 2000, see Eric Katz, "Understanding Moral Limits in the Duality of Artifacts and Nature: A Reply to Critics," *Ethics and the Environment* 7 (2002: 138–146. While in some respects Katz now admits to the utility of restoration in some cases, as I argue in a short commentary in the same issue of this journal, his ultimate critique of restoration and his concerns about it have not changed. He argues here, for example, that "once we begin to adopt a general policy of remediation and restoration, we may come to feel omnipotent in the manipulation and management of nature. And thus we will create for ourselves a totally artifactual world" (142).
2. See Andrew Light and Eric Higgs, "The Politics of Ecological Restoration," *Environmental Ethics* 18 (1996): 227–247.
3. While I will not go into the details of my own metaethical views in environmental ethics, I have termed them "methodological environmental pragmatism." For explanations of what I mean by this, see my contributions to Andrew Light and Eric Katz, eds., *Environmental Pragmatism* (London: Routledge, 1996), as well

as my essay "The Case for a Practical Pluralism," in Andrew Light and Holmes Rolston III, eds., *Environmental Ethics: An Anthology* (Cambridge, Mass.: Blackwell, 2003), pp. 229–247; and "Taking Environmental Ethics Public," in David Schmidtz and Elizabeth Willott, eds., *Environmental Ethics: What Really Matters? What Really Works?* (Oxford: Oxford University Press, 2002), pp. 556–566. I do not believe, however, that the arguments I present here rely on this framework, though one can no doubt trace the rationale for my concerns over the philosophical literature on restoration ecology to the general policy orientation of that other part of my work.

4. Katz, *Nature as Subject*, p. 97 (originally in "The Big Lie," as are KR2–KR4). KR1 is restated later in Katz, "The Call of the Wild": "what makes value in the artifactually restored natural environment questionable is its ostensible claim to be the original" (p. 114).
5. Ibid., p. 97.
6. Ibid., p. 97. KR3 is most thoroughly elaborated later in Katz, "Artifacts and Functions."
7. Ibid., p. 105. The domination argument is repeated in Katz, "The Call of the Wild," p. 115. The argument is also repeated in Katz, "Artifacts and Functions," and further specified in idem, "Imperialism and Environmentalism." As far as I can tell, though, the argument for domination is not really expanded in this last paper, except that imperialism is deemed wrong because it makes nature into an artifact (KR3).
8. Ibid., p. 113, from the chapter "The Call of the Wild" and repeated in idem, "Imperialism and Environmentalism," p. 139.
9. See Katz, "Understanding Moral Limits."
10. The way I have put this distinction previously is that a benevolent restoration, unlike a malicious restoration, cannot serve as a justification for the conditions that would warrant its engagement. I make this distinction to avoid the kinds of cases that Robert Elliot has considered in the past where a firm argues that a form of development should be permitted because whatever bit of nature will be destroyed to engage in some form of natural resource extraction can always be restored later. I believe that such cases can be independently answered without recourse to this distinction, though it is helpful to clarify the kinds of restorations that are largely unproblematic. Interestingly enough, the Society for Ecological Restoration, the primary organization that sets priorities and standards for restorations, specifically prohibits malicious restorations as its first policy recommendation to its membership.
11. Katz, "Understanding Moral Limits," p. 105.
12. Ibid., p. 105.
13. Just as, on a broader scale, there can be a town full of decent, law-abiding citizens that may not constitute a moral community in any significant sense.
14. It is also the case that restoration will only be one out of a large collection of practices available for adaptive management. In a project to clean up an abandoned mine site, for example, we can imagine a case where restoring the site to

a landscape that was there before would not be the best choice and that instead some other sustainable landscape that would help to preserve an endangered species now in the area would be more appropriate. I am indebted to Anne Chapman for pressing me to clarify this point.

15. Katz can legitimately respond here that there seems to be no unique reason why people couldn't get these kinds of experiences that generate a closer relationship with nature out of some other kinds of activities. Why couldn't we just use this sort of argument to encourage more acts of preservation, or simply more walks though nature, or the like? Such an objection would, however, miss a crucial point. Even if it can be proved that we can get these kinds of positive experiences with nature through means other than acts of restoration (and I see no reason why we couldn't), this does not diminish the case being built here: that restoration does not necessarily result in the domination of nature. Arguably, moreover, our experiences as restorationists give us some of the kinds of understandings of the workings of natural processes required for aesthetic appreciation. (See Allen Carlson, "Nature, Aesthetic Appreciation, and Knowledge," *Journal of Aesthetics and Art Criticism* 53 [1995]: 393–400.) Importantly, this understanding is a transitive property: it gives us an ability to appreciate aesthetically not only the nature we are trying to restore but also the nature we are not trying to restore. Restoration thus could provide a unique avenue into the aesthetic appreciation of all nature, restored or not.

16. This paper is a shortened version of my "Ecological Restoration and the Culture of Nature: A Pragmatic Perspective," in Paul Gobster and Bruce Hall, eds., *Restoring Nature: Perspectives from the Social Sciences and Humanities* (Washington, D.C.: Island, 2000), pp. 49–70, slightly modified for this volume. Further, regarding my thinking on this topic, see my "Restorative Relationships," forthcoming in Robert France, ed., *Healing Nature, Repairing Relationships: Landscape Architecture and the Restoration of Ecological Spaces* (Cambridge, Mass.: MIT Press, 2005).

TEN

Ecological Restoration and the Renewal of Wildness and Freedom

MARK WOODS

THE EVERGLADES National Park Act of 1934 mandated an area of south Florida to be "permanently reserved as a wilderness," where the flora and fauna were to be preserved intact in "essential primitive conditions."[1] Today, nearly 1.3 million acres of the 1.5 million acre Everglades National Park are managed as the Marjory Stoneman Douglass Wilderness Area (MSDWA) within the National Wilderness Preservation System of the United States. The Everglades is the largest freshwater marsh in the world and contains seven distinct ecological matrixes of sawgrass everglades, mangrove forests, salt marshes, cypress forests, pine forests, West Indian hardwood hammock forests, and coastal prairies. What is protected as wilderness, however, seems too small to survive. The national park contains the remnants of a larger freshwater region of the Everglades ecologically dependent on a continuous, flushing supply of water. Before European and Euramerican settlement of the region, this supply began in a series of lakes two hundred miles to the north of the national park that created a flow of water south into Lake Kissimmee and the Kissimmee River Basin, into Lake Okeechobee farther south, and eventually into what is now the Everglades National Park. Beginning in 1907 and continuing today, virtually all rivers and streams in south Florida have been converted into canals, many of which now drain east, to the Atlantic coast. A dike has been constructed around the southern edge of Lake Okeechobee, and the flow of freshwater into the Everglades National Park has been reduced by up to 90 percent. The remaining 10 percent of the historic water flow, however, can come almost all at once during extremely wet periods when water stored in agricultural and urban areas to the north and northeast of the park is released to prevent flooding outside the park. As a result of water diversion projects in south Florida, either too little or too much freshwater reaches the MSDWA to sustain its flora and fauna.

In order to right this ecological wrong, several Florida state and U.S. federal agencies have been mandated to restore the Everglades National Park. Many of their efforts have been directed toward controlling pesticide and fertilizer runoff pollution from agricultural areas between Lake Okeechobee and the national park (Holloway 2000). In January 1997 government officials and conservationists officially broke ground with golden shovels to begin an ambitious engineering attempt to alter and increase the flow of water into the national park. This engineering attempt is an act of ecological restoration and, if successful, will restore some of the needed flow of water back into the area.

Justifying this act of ecological restoration raises what William Throop (1997) has called a "paradox of preservation." We begin with the belief that the restoration of the Everglades is preferable to (1) further human alteration of the area and to (2) leaving the area alone and letting it evolve on its own, following from current anthropogenic alterations. We might justify the restoration of the Everglades by appealing to the natural values of the area and the fact that certain historical processes—continuous clean water flows—need to be restored to ensure the continued existence of the Everglades as a natural area. But an important, if not essential, characteristic of naturalness is that what is natural has come into being independent of people (Elliot 1997),[2] and natural processes supposedly pursue their own independent and unplanned courses of development (Katz 1997). If ecological restoration is warranted, haven't natural values already disappeared, and how can something distinctively nonhuman—naturalness—be restored by human actions?

Ecological Restoration and Time

Temporality is an important consideration for ecological restoration: to restore something seems by many to be an attempt to set back time and return an area to some previous condition. Restoration efforts in the Everglades revolve around restoring a clean, flowing supply of water to the area. If we define the term "restoration" as literally meaning "to bring back to an original state through an act of ecological repair," the restoration of the Everglades remains problematic. The original state of the Everglades (more than a hundred years ago) contained a flow of clean water uninterrupted by diversion canals and dikes, and it is misguided to think that we are going to remove every human structure that diverted water away from the Everglades and restore the exact water that once flowed into the area. Rather, what is intended here is a return to a historical pattern of the flow of water from the Green Swamp in central Florida south to Florida Bay. Through various efforts, it might seem plausible that the pattern and its volume and frequency can be restored.

More than this, however, is intended. Wading bird populations in the Everglades have declined by about 90 percent in the past century, as have populations of many of the area's native animals and fish. Restoring the supply of clean water to the area is thought to be a necessary first step in restoring historical, native populations of organisms. Obviously, dead organisms and extinct species cannot be brought back to life, but perhaps viable populations of snail kites, wood storks, crocodiles, panthers, and other organisms that are in danger of extinction can continue to exist here. We are attempting to restore the landscapes and distinct ecosystems of the Everglades back to the condition they were in roughly a hundred years ago before the area was significantly colonized and developed by Europeans and Euramericans.

Given this temporal goal of the ecological restoration of the Everglades, four objections can be raised. First, there are incredible practical difficulties in any act of restoration, and we lack precise ecological, biological, and hydrological knowledge of the "primitive conditions" (as stipulated by the Everglades Park Act of 1934) of this area of Florida before the twentieth century. Today, large gaps in our knowledge of this area still exist, and there may always remain doubt about whether our restoration efforts are historically accurate. Because of our lack of knowledge about the past and because of uncertainty of the future, some, such as Eugene Hargrove (1989), have articulated a form of "therapeutic nihilism" (158–61), whereby we do nothing to restore an area of nature.[3]

Even if we had good ecological, biological, and hydrological knowledge of the Everglades, however, a second objection remains: nature does not stand still in some form of equilibrium that can be restored (Pickett and White 1985; Botkin 1990). Given that natural landscapes are constantly changing in many ways, by the time we restore a natural landscape, we may have no idea what the current condition would have been without our influence. To turn back the clock to the point at which the landscape was denuded or destroyed may deny natural change.

A third objection follows: what is the original "primitive condition" of the Everglades? It seems arbitrary to pick one presettlement point in time—roughly one hundred years ago in the case of the Everglades—and restore the conditions that existed then. To do so enshrines a freeze-frame mentality (Callicott 1991; Callicott 1994/95) of ecological restoration that arbitrarily selects a point in time that we act to restore.[4]

Fourth, people certainly affected the Everglades area earlier than a hundred years ago. Paleo-Indians occupied and changed parts of the area thousands of years ago. Now-extinct groups of Calusas and Tequestras were also here more than a hundred years ago, as were modern groups of Miccosukees and Seminoles. The consensus today is that the impacts of Indians were significant (Denevan 1992; Cronon 1995), and we cannot claim that all of Florida existed in some condition of wilderness empty of people before the arrival of Europeans and Euramericans. Most acts of ecological restoration aim to restore an area to a condition before human impacts, but human impacts for thousands of years in Florida—many of which we are just beginning to understand, some of which we may never understand—make the restoration of the Everglades, to a condition they were in before human impacts, problematic.

The Restoration of Naturalness?

Given the above problems, should the goal of ecological restoration be to return the Everglades to a previous temporal condition? Many people might wish to claim that the previous temporal condition of the Everglades was a natural condition and that ecological restoration should be aimed at restoring naturalness.[5] But what is naturalness? Robert Elliot argues that naturalness is a value-adding relational property that connotes that something is "unmodified by human activity" (1997:82). The significance of naturalness for Elliot stems from its nonhuman genesis and causal continuity with the past. Other philosophers have made similar claims about naturalness. Paul Taylor defines the term "natural" in terms of "biological and environmental factors" that "take place without human intervention" (1986:4). Holmes Rolston defines the natural as "spontaneous nature" that "runs by causal law, biological metabolisms, genetic coding, instincts, evolutionary and ecological processes, [and] accidental contingencies" (1994:4). Robert Goodin's (1992) green theory of value is grounded in a process view of the nonhuman history of naturalness. Eugene Hargrove's (1989) account of nature as a "self-created otherness," in contrast to humans, identifies naturalness as a key property of nature. Following these accounts, we can understand naturalness to be a relational property that refers to a causal history characterized by nonintentional, ateleological physical, chemical, biological, and evolutionary forces (Kramer and Woods 2003). A central problem that emerges for ecological restoration is that naturalness itself cannot be restored directly by human actions that are the result of intentional, cultural forces. Restoring natural, historical patterns of water flows in the Everglades seems impossible because naturalness is anathema to direct human intervention.

One way around the problem of restoring naturalness is to problematize the account of it I have sketched above. There are at least two central objections: (1) The manner in which I have defined naturalness is too stringent: —because of persistent human impacts everywhere on the planet, there already are no natural areas left, so why should we worry about restoring what has long since disappeared? (2) This account of naturalness necessitates separating human cultures from nonhuman natural nature, but humans and their cultures are entirely natural. Let me briefly suggest why these objections fail and naturalness remains a problem for ecological restoration.

First, the account of naturalness sketched above seems to enshrine a purity definition of naturalness: to qualify as natural, an area must have remained unaffected (pure) of human impacts in the past and must be so unaffected today.[6] If we follow such a definition, one person who walked across dry land in the Everglades two hundred years ago might have destroyed the area's naturalness. Given past and present human impacts all over the planet everywhere except Antarctica for at least thousands of years (and including Antarctica today) (McKibben 1989; Callicott 1991), we might be hard-pressed to find any naturalness in the Everglades or anywhere else today. But we should be careful here when discussing human impacts because all human impacts are not of the same magnitude. There is a considerable difference between a set of footprints on a sandbar at Flamingo and the straightening of the Kissimmee River into a fifty-six-mile canal called C-38. The sandbar, although affected by people, might still remain largely natural. In contrast, the Kissimmee River has been turned into a human artifact that is no longer a natural river. Other features of the Everglades might be more or less natural to the degree that they have or have not been modified by intentional human activity. This suggests that naturalness and its opposite—artifactuality—exist along a continuum and do not admit of necessarily sharp metaphysical boundaries (Elliot 1997; Kramer and Woods 2003), and we should reject a purity definition of naturalness.

But this account of naturalness seems to require a strong boundary between the human and nonhuman worlds, and a number of people have claimed that *Homo sapiens*, like all other species on Earth, has evolved from common ancestors and is an entirely natural species (Callicott 1991; Callicott 1994/95; Turner 1994; Ashworth 1999). Although a full discussion of the human culture/nonhuman nature distinction warrants more space than I can devote to it here, a number of people have argued that there are good reasons to retain a culture/nature distinction. Max Oelschlaeger (1991:8) argues that an understanding of quintessential nonhuman nature—wilderness—is necessary for us to recognize ourselves as cultured beings.[7] Kate Soper (1995:38–39) argues similarly that much of our human discourse presupposes a distinction between ourselves and nonhuman nature as a precondition for meaningful articulation of such discourse. Both Elliot (1997) and Rolston (1994) argue that, although we evolved from ancestors common to other natural organisms and were natural ourselves in the past, our strong degree of separation from nonhuman nature today, which includes going beyond strict biological and ecological laws of evolutionary fitness, has made us more or less cultural beings as opposed to natural beings.[8] I have argued elsewhere (Woods 2001) that these differences make a culture/nature distinction meaningful when contrasting human cultures and natural wilderness.

If these arguments for culture/nature distinction are cogent, and if naturalness still exists in various degrees today, then naturalness remains a problem for ecological restoration. The flow of water through the Everglades is largely not natural today because people (cultural forces) intentionally control and shape what that flow is. If the ecological restoration of the Everglades is successful, we will change this flow to something resembling a historical pattern. But this re-created pattern still will be largely shaped by us, and the flow—like most other water processes in Florida—will have to be managed to be maintained. Our ability to create or re-create natural processes, objects, or areas remains questionable because our actions and their consequences cannot escape a cultural context that shapes the what, how, and why of our actions (Moriarty and Woods 1997). Some significant degree of nonhuman genesis and continuous causal history seems to be necessary for naturalness, without which the natural becomes more like the artifactual fashioned by people. Because ecological restoration involves a conscious manipulation of nature by people, and because by the time restoration is warranted the natural history of an area has already been interrupted by cultural forces, we should remain skeptical of our ability to restore naturalness.[9]

Wildness and Freedom

If the restoration of naturalness in the Everglades is problematic, what else could we be attempting to restore? Many claim that the ecosystems of the Everglades are unhealthy because they do not receive enough clean water, and the notion of ecosystem health has received much attention in recent years (Costanza, Norton, and Haskell 1992; Grumbine 1992, 1994).[10] Because the Everglades now lacks a number of processes, organisms, and species that once maintained it, some might argue that what needs to be restored is the integrity of the area (Westra 1994). Biodiversity has been reduced in the area in the past one hundred years, and we could be seeking to restore the area's biodiversity by restoring the water flows.[11] The balance (Pimm 1991) of the area has been disrupted, as have its resilience (Walker 1995) and fragility (Nilsson and Grelsson 1995), and restoration efforts could be aimed at these. Other values such as the Everglades' ecological complexity and richness also could be at stake in restoration efforts.[12]

If the goal is to restore something such as the health, integrity, or biodiversity of the Everglades, most restoration efforts aim at particular kinds of health, integrity, and biodiversity that exist largely independent of people. There is much biodiversity at the nearby Metro Zoo in Miami, but the flora and fauna that compose it have been gathered and are extensively managed by people. This is not what most people want for the Everglades. Instead, those who seek to conserve biodiversity in the Everglades seek to conserve natural biodiversity. But the restoration of natural biodiversity is problematic. We could, instead, consider wild and free biodiversity.

> Would you be satisfied if most species on Earth were maintained in minimally viable populations, but were all securely confined to zoos, aquaria, botanical gardens, and the like? I have not found any conservation biologists who are comfortable with this level of domestication. Most conservation biologists, implicitly or explicitly, are talking about maintaining wild populations in wild environments. Truly wild. A landscape where every acre is intensively managed (increasingly the case in National Forests, for example) is not much different from a zoo. I do not believe that the containment of biodiversity in zoos—indoor or outdoor—provides a satisfying vision or a sustainable mission for conservation biology.
>
> (NOSS 1995/96:19)

Ecological restoration efforts in the Everglades largely are directed by conservation biologists who seek to restore biodiversity, ecosystem health, integrity, and so on that exist independent of civilizing and confining forces. Reed Noss admonishes us to direct our attention to ecological processes and assemblages that are beyond civilizing controls and confinements: —that are wild and free. As a first step toward protecting wildness and freedom in the Everglades, restoration efforts might be directed toward renewing wild and free patterns of water flow.

But what does it mean for water flows and ecological processes and assemblages to be wild and free? Consider first wildness. Wildness, like naturalness, typically is predicated of nonhuman nature. Given human impacts virtually everywhere on Earth, what is wild will be more or less wild, just as what is natural will be more or less natural. Wildness, however, should not be equated with naturalness. Accounts of wildness focus on what we might call the autonomy or authenticity of nature. Eric Katz claims that natural processes "pursue their own independent and unplanned course of development" (1997:115). Andrew Brennan (1984) claims that natural objects were not consciously designed to have functions either in nature or to people and thus lack what he calls any "intrinsic function"; this lack of an intrinsic function implies that these objects can pursue their own course of development. Keekok Lee develops a similar claim out of biological evolutionary theory: nonhuman nature has no teleological direction, and because Earth and its biosphere can exist independently of people, "the Earth and its extremely complex biosphere are fully autonomous" (1994:93). Thomas Birch gives an account of wildness in terms of the otherness "that permits sheer spontaneity and continuous participation in the emergence of novelty" (1990:11). Neil Evernden (1992) argues that wildness is nature that exists for itself independent of human designs. Chris Cuomo claims that a living being exhibits "*dynamic charm*—its diffuse, 'internal' ability to adapt to or resist change, and its unique causal and motivational patterns and character" (1998:71). Given these accounts of wildness, we can begin to understand it as the autonomy of the more-than-human world where events such as animals moving about, plants growing, and rocks falling occur largely because of their own internal self-expression that is independent of civilized forces.[13]

Immediately we are confronted with an objection: this stretches a common philosophical understanding of autonomy too far. As famously put forth by Immanuel Kant ([1785] 1964), a central feature of autonomy is that one (usually a human) must rationally legislate for oneself by willing independently of outside influences. I autonomously choose to write this paper, but it might seem difficult to ascribe a similar degree of autonomy to environmental entities that lack wills, senses of self, or the abilities to deliberate rationally. Yet we can understand movements and happenings in nature that are not willed in a humanized, civilized, and rational sense but occur spontaneously and independently of human actors. Evolution occurs over billions of years; organisms move around, eat, grow, reproduce, and die; lightning strikes, and vegetation ignites in fire. Movements and happenings occur independent of and in spite of human intentionality, and wildness denotes the autonomy of the more-than-human world. Similar to autonomy, wildness disappears when entities and forces are reshaped to meet the imposed expectations of others' (civilized peoples') wills (Kramer and Woods 2003). Consider a wild horse that becomes tamed. Before taming, the horse acts in conformity with its own spontaneous, self-directed will and remains authentic to itself as a wild horse. After taming, the horse internalizes the controlling wills of people, redirects its own will in conformity with these other wills, and is no longer authentic to itself.

Although a number of people describe wildness as a form of freedom (Brennan 1984; Birch 1990; Hettinger and Throop 1999), freedom importantly is distinct from wildness (Kramer and Woods 2003).[14] Where wildness connotes a self-expressed form of autonomy, freedom connotes having a broad range of opportunities for self-expression that is unhindered by confining, external forces. That is, what is wild becomes civilized to the degree that it internalizes civilized and other-willed forces, and what is free becomes confined or imprisoned to the degree that external constraints force it to conform to prescribed limits. In his article "The Incarceration of Wildness: Wilderness Areas as Prisons," Birch (1990) is concerned about how the wildness of wilderness is tamed and brought under control, but he also addresses the loss of wilderness freedom when external limits are placed on wilderness areas. Hettinger and Throop define wildness in terms of not being humanized, but they speak about humanization as the degree to which something is "influenced, altered, or controlled by humans" (1999:12); this speaks to both the civilizing of wildness and the confining of freedom. Capturing and caging animals, restricting the range of plants and animals through habitat destruction, and damming rivers are all examples of confining freedom, as is cutting off historical patterns of water flows in the Everglades.

Articulating wildness and freedom in these terms reveals an important difference between these properties and naturalness. When natural history has been broken, the natural has been replaced with the artifactual. Likewise, when the autonomy of wildness has been brought under control, the wild has been civilized as the tame, and freedom is replaced with confinement and imprisonment when external constraints force something into prescribed limits. Because a central characteristic of naturalness is its nonhuman genesis and history, the return of naturalness seems problematic in the near future. It might be only over very long timescales that artifactuality might wash out of an area as naturalness slowly washes back into it. In contrast, wildness and freedom might return rather quickly to the extent that people and cultural forces are no longer exerting control over an area. As an illustration of this, consider the following example. Along one side of the road that runs between the Shark Valley Information Center and the Shark Valley Observation Tower on the northern boundary of the Everglades National Park, one can see extensive cattails growing in a canal dredged for the road. The cattails are not native to this area of the Everglades. They and the canal have a history shaped by human intrusion into this area in the 1940s, and, because of this, they may exhibit little naturalness. In contrast, the cattails resist our best eradication efforts, and the canal area contains populations of alligators, wading birds, fish, and other organisms that exhibit some degree of wild autonomy and freedom. When the road was first constructed, naturalness, wildness, and freedom were diminished, but wildness and freedom seem to have reemerged. Freedom might have come back rather quickly to the degree that external humanizing constraints were removed or disappeared. Wildness also might have reemerged rather quickly. For organisms such as the animals living near the road, wild autonomy might have reemerged in as little as one or two succeeding generations. But if a characteristic feature of naturalness is its causal history independent of intentional human cultural forces, the naturalness of Shark Valley seems to be diminished for a significantly longer timescale.

Ecological Restoration as the Renewal of Wildness and Freedom

Because wildness and freedom can reemerge over relatively short timescales, I believe that the goal of ecological restoration should be the restoration of wildness and freedom. But isn't ecological restoration itself an instance of deliberately fashioning nature in accordance with human designs? If so, it might be anathema to both the wild autonomy and freedom of the more-than-human world. In "The Big Lie: The Human Restoration of Nature," Eric Katz (1992) argues precisely this. Katz claims that ecological restoration cannot restore wild nature because wildness quintessentially exists outside of human control, and deliberative human action (an instance of control) is anathema to wildness.[15] Because wild nature exists for itself, independent of people, and lacks intrinsic functions, it lacks any blueprints that we might use as guides for ecological restoration. Ecological restoration is an instance of the human trammeling of wild nature because nature is not permitted to be autonomous and is subjected to further human controls via restoration.[16] If this is true, how can we justify restoring wildness and freedom in the Everglades?

Remember that we began with the belief that the restoration of the Everglades is preferable to (1) continuing to alter negatively (in an ecological sense) the area through anthropogenic impacts and to (2) simply leaving the area alone. If we continue to alter the area negatively(1), we continue to extinguish the wild autonomy and confine the freedom of the area. If my accounts of wildness and freedom are accurate, wildness and freedom can return to the Everglades in the absence of our control and continued alteration of the area, that is, simply by removing ourselves from the area and leaving it alone (2). Thus ecological restoration efforts are unnecessary. Further, such efforts may continue to extinguish wildness and freedom by subjecting the area to further anthropogenic impacts—Katz's objection to restoration.

I suggest that we can begin to circumvent these objections to ecological restoration by appealing to the claim that wildness and freedom can come in degrees: an area can be more or less wild and more or less free.[17] At least some wildness and freedom now exist in the Everglades. More wildness and freedom might exist here fifty years from now if we were simply to remove anthropogenic obstacles to a flow of clean water through the area. If various restoration efforts will plausibly result in even more wild autonomy and freedom in the area fifty years from now, they may be justified. For example, with no further restoration efforts, panthers most likely will be extinct from the Everglades in the near future. Even with current restoration efforts, they may still go extinct, but the probability of this may be less than if no restoration efforts take place.[18] The area is more wild and free with panthers than without them. We can view panther restoration as a rehabilitation effort to help keep wild and free products (the panthers) and wild and free processes (predation) in an area that is wildly enriched with them. The goal of ecological restoration can be seen in terms of helping create future conditions whereby wildness and freedom will be enriched.[19] Thus ecological restoration can be preferable to merely leaving an area such as the Everglades alone. The end goal is future-oriented wildness and freedom understood as the autonomy and unconfined expression of the more-than-human world. In contrast to this sense of diachronic wildness and freedom, Katz's objection to ecological restoration is that it trammels currently existing wildness and freedom: synchronic wildness and freedom. If restoration efforts are warranted, chances are good that synchronic wildness and freedom presently are being trammeled. If my accounts of wildness and freedom are correct, Katz cannot claim that restoration trammels diachronic wildness and freedom because such wildness and freedom can reemerge both with and without restoration. And if more diachronic wildness and freedom might emerge with restoration efforts, then such efforts might be justified.

I wish to point out that the renewal of diachronic wildness and freedom via ecological restoration efforts need not be confined to wilderness settings such as the Florida Everglades. Although this is my focus in this paper, wild and free animals, plants, species, landscapes, and so on can be found in both environments that are less humanized—such as wilderness areas—and in environments that are more humanized—such as farmlands, urban greenbelts, and city parks. While there can be a strong temptation to equate wildness and freedom with wilderness, we should resist this temptation (Turner 1996). Wilderness certainly may be a significant location of wildness and freedom (as well as naturalness), and respecting its wildness and freedom (and naturalness) probably means keeping wilderness free of humanization as much as possible, that is, wilderness preservation. But respecting wildness and freedom outside of wilderness areas might even allow some form of human participation with nature, as Ned Hettinger argues in this volume.[20] Hettinger makes a distinction between the control or mastery of nature and the modification of nature in a less intrusive manner. I have made a similar distinction elsewhere between what I call trammeling nature and impacting nature (Woods 2001). In this volume, Eric Katz denies this distinction and categorizes all human impacts as a form of the control or domination or nature,[21] and this is why ecological restoration is problematic for him. But ecological restoration efforts can be respectful of wildness and freedom (what Hettinger calls nature's autonomy) to the degree that such efforts do not lead to the control, mastery, or trammeling of nature.

Another way to respond to Katz's objection that ecological restoration, like ecological destruction, is just another form of the human domination of nature is to point out that there are many different types of ecological restorations. In his article "Mucking with Nature," Richard Sylvan (1994) draws distinctions between the environmental pillage, reclamation, and partial reclamation (Sylvan calls this last category "extitution") of substantially natural areas and the rehabilitation, reinhabitation, resettlement, and creation of parks and gardens in damaged or modified areas. When a mining company wishes to destroy a natural area through mining activities and attempts to justify this by promising to restore the area after the mining, this case of extitution might be a fair target for Katz's criticism that restoration is being used to justify the domination of nature. If, however, we rehabilitate a formerly natural area that has been denuded by mining activities, where there was no pretense that the mining was justified by some future restoration efforts, we may not be dominating nature but instead merely trying to rehabilitate a denuded area. Andrew Light (in this volume)[22] calls this "benevolent restoration," as opposed to the "malicious restoration" efforts Katz addresses, and argues that restoration efforts can actually free nature from humanized constraints.[23]

While viewing restoration in a benevolent manner responds directly to Katz's criticism of restoration, his criticism should not be dismissed altogether. Many restoration efforts can be based on a domination model. A number of people view restoration ecology—the science of ecological restoration—as an applied form of ecology that is undertaken for the heuristic purpose of learning "how nature works" (Cowell 1993:20). Although restoration ecologists typically seek to restore nature, they also typically seek to understand how nature works by actively manipulating it. As John Harper (1987) argues:

> The *raison d'être* for a science of ecology is presumably the development of an understanding of the workings of nature that would enable us to predict its behavior, and to manage and control (conserve or change) it to our liking.... I have argued elsewhere that there is an analogy between the study of ecological phenomena and the study of clocks and watches, and this analogy is useful here.... To discover how a watch works, the repairer does not simply describe it, but takes it to pieces and puts it back together and, if it works, gains at least some understanding of the way in which it does so. The point is that this understanding is achieved by *doing* something to the watch in an attempt to

> get it going again. . . . To restore such systems [communities and ecosystems] successfully he or she must understand how they work. Conversely, perturbing the system, taking it apart and trying to reassemble it, is one of the best ways to achieve this understanding. Restoration ecology is just this process of assembly, carried out specifically to test ideas about how communities and ecosystems work.
>
> (35, 36, 38)

Harper's analogy of restoration ecology and watch repair is telling. There are two dangers here. First, nature becomes a "tinkering laboratory" for the restoration ecologist whose goal is scientific understanding. Second, the goal of scientific understanding seems to lead to the control of nature, and the control of nature is anathema to wildness and freedom. When Frederick Turner (1994) views restored landscapes as "invented landscapes" actively created and maintained by intentional human actions much as a gardener continually maintains a garden, can wildness and freedom reemerge? When William Jordan (1994) argues that restoration allows us to participate continually as civilized beings with nonhuman nature, is there room for wildness and freedom in a more-than-human world? When officials from the Southwest Florida Water Management District—the state agency that manages all public water in the Green Swamp and other areas north of the Everglades where the flows of water that flush the Everglades begin—state that few lightning-caused fires will be allowed to burn because of lack of controls and prior planning and instead lightning fires are now largely replaced with controlled burns, have wildness and freedom been significantly diminished? The worry that restoration ecology raises for my account of ecological restoration is that wildness and freedom—and perhaps naturalness across longer timescales—are not restored but instead might be continually replaced with civilized and confining forces. If restoration ecology as a discipline is to be of help in restoring wildness and freedom, restoration ecologists must reconceive their goals of manipulation, management, and control of nature in the name of the advancement of ecological knowledge and recognize that their work is temporary and should attempt to renew wildness and freedom.[24] As Rolston reminds us: "Any restoration is an artifact at the moment that it is deliberately arranged, but it gradually ceases to be so as spontaneous nature returns—but, if, *and only if*, humans back off and let nature take its course" (1994:92; italics added).[25]

Ecological restoration properly should be viewed as efforts to help renew wildness and freedom in an area. The term "renewal" implies a revival or a regeneration and might express better the idea that we wish to help renew diminished values of wildness and freedom. We may be hard-pressed to restore something such as naturalness, and it might be the case that nature can accomplish this without us over long timescales. Over the short term, however, we may be able to help renew diachronic wildness and freedom in trammeled areas such as the Everglades.

Conclusion

Restoring the Everglades to a nonanthropogenic historical pattern (continuous clean flows of water) seemed to involve restoring the area to a specified point in time—so-called presettlement—when this pattern existed. This, however, was problematic for four reasons:

1. We lack precise ecological knowledge of the area before anthropogenic disturbances.
2. Nonanthropogenic historical patterns change, and to set back the clock through restoration is to deny this change.
3. It seems arbitrary to pick one previous point in time as the goal of restoration.
4. Reinforcing point 1, the area has been impacted for thousands of years by people.

If we envision the goal of ecological restoration to be future-oriented in terms of the renewal of diachronic wildness and freedom in the Everglades, we can begin circumventing these backward-looking objections to restoration.[26] To the degree that we remove civilizing forces and keep such forces out of the Everglades, wildness can be renewed, and to the degree that we remove external constraints and limitations imposed by cultural forces in the Everglades, freedom can be renewed. The restoration efforts begun in 1997 to renew water flow patterns can be a significant step toward a wild, free, and—in the distant future—natural Everglades.

NOTES

1. An act to provide for the establishment of the Everglades National Park in the state of Florida and for other purposes, approved May 30, 1934 (48 *Stat.* 816).
2. See also Elliot (1982).
3. Support for therapeutic nihilism can be found in Leopold's admonishment that because we lack so little ecological knowledge we should know that to "keep every cog and wheel is the first precaution of intelligent tinkering" (1966:190) and in Commoner's third law of ecology that "Nature knows best" and that "any

man-made change in a natural system is likely to be *detrimental* to that system" (1971:41).

4. Botkin (1990:58–59) raises a similar problem for the Boundary Waters Canoe Area in Minnesota.
5. For example, see Cowell (1993).
6. See Foote (1973) for a similar account and rejection of a purity definition of wilderness.
7. See also Oelschlaeger (1999) for an account of why human culture should not be conflated with nonhuman nature.
8. Who we are as cultural beings supervenes on who we are as biological and ecological beings.
9. Anderson (1991) attempts to define naturalness in terms of the degree to which a landscape would not change if humans were removed. Anderson proposes gauging naturalness by measuring both "the complement of native species currently in an area compared with the suite of species in that area prior to settlement" and "the amount of cultural energy, the energy subsidy supplied by technological man [*sic*], required to maintain the functioning of the system as it currently exists" (p. 348). As discussed above, "prior to settlement" raises problems. Although we tend to think of wilderness areas as places we generally leave alone (where we expend little if any "cultural energy"), the restoration of the Everglades seems to require large amounts of "cultural energy."
10. See also the special issue of *Environmental Values* 4 (1995) devoted to ecosystem health.
11. See Wilson (1992) for an account of biodiversity and Noss and Cooperrider (1994) for an account of restoring biodiversity.
12. See Rolston (1994) for an account of many of these different kinds of ecological values.
13. I am borrowing the phrase "more-than-human world" from Abram (1996).
14. The reintroduced wolves in Yellowstone National Park are probably wild, but, because they have been tracked with radio collars and are killed or relocated if they stray outside of certain prescribed areas, they are not free. In contrast, a domesticated dog or cat that is no longer wild but escapes its confinement in a house or yard is now free. I lack the space here to discuss a more complete theory of the properties of wildness, freedom, and naturalness and to discuss the distinctions and similarities of these three properties; see Kramer and Woods (2003) for a more complete account of such a theory.
15. Katz specifically addresses wildness and does not make a distinction between wildness and freedom; however, he alludes to freedom and in a number of places runs together the distinction I make between wildness and freedom.
16. See also Katz (1997).
17. Consistent with this, Bratton (1992) reminds us that most environments lie somewhere between fully pristine, natural landscapes and fully artifactual, humanized landscapes.
18. See Shrader-Frechette and McCoy (1993) for an extended philosophical discus-

sion of efforts to save the Florida panther.

19. More than likely, there will be some limits set by other values to the enrichment or maximization of wildness and freedom. Eliminating ourselves from the planet might bode very well for enriching the wildness and freedom of nonhuman nature, but this would be overridden by various cultural values wrapped up in human survival and flourishing.
20. "Respecting Nature's Autonomy in Relationship with Humanity."
21. "The Liberation of Humanity and Nature." A denial of this distinction is present in most of Katz's work.
22. "Restoration, Autonomy, and Domination: The Case for Benevolent Restoration."
23. Lo (1999) argues similarly that restoration need not be dominating as Katz says and instead can be a way for people to cooperate with nature and respect its autonomy.
24. Although they do not speak about renewing wildness and freedom, Jackson, Lopoukhine, and Hillyard (1995) claim that ecological restoration should repair damage caused by humans and should concentrate on restoring ecological and biological processes that direct nature. To the degree that these processes can return, nature can again be autonomous and free.
25. See also John Sandlos's essay in this volume ("Purple Loosestrife and the 'Bounding' of Nature in North American Wetlands") for an extended discussion of how restoration efforts that involve the eradication of purple loosestrife in wetlands can be viewed as problematic in a similar sense as the continued human control of nature.
26. See Rolston's (1994:93) account of restoration, where he argues that restoration should be viewed as a forward-looking rather than backward-looking event.

REFERENCES

Abram, David. 1996. *The Spell of the Sensuous: Perception and Language in a More-Than-Human World*. New York: Pantheon.

Anderson, Jay E. 1991. "A Conceptual Framework for Evaluating and Quantifying Naturalness." *Conservation Biology* 5: 347–352.

Ashworth, William. 1999. *The Left Hand of Eden: Meditations on Nature and Human Nature*. Corvallis: Oregon State University Press.

Baldwin, A. Dwight Jr., Judith De Luce, and Carl Pletsch, eds. 1994. *Beyond Preservation: Restoring and Inventing Landscapes*. Minneapolis: University of Minnesota Press.

Birch, Thomas H. 1990. "The Incarceration of Wildness: Wilderness Areas as Prisons." *Environmental Ethics* 12: 3–26.

Botkin, Daniel B. 1990. *Discordant Harmonies: A New Ecology for the Twenty-First Century*. Oxford: Oxford University Press.

Bratton, Susan Power. 1992. "Alternative Models of Ecosystem Restoration." In *Ecosystem Health: New Goals for Environmental Management*. Ed. Robert Costanza, Bryan G. Norton, and Benjamin D. Haskell. Washington, D.C.: Island.

Brennan, Andrew. 1984. "The Moral Standing of Natural Objects." *Environmental Ethics* 6: 35–56.

Callicott, J. Baird. 1991. "The Wilderness Idea Revisited: The Sustainable Development Alternative." *Environmental Professional* 13: 235–247.

——. 1994/95. "A Critique of and an Alternative to the Wilderness Idea." *Wild Earth* 4: 54–59.

Commoner, Barry. 1971. *The Closing Circle: Nature, Man, and Technology*. New York: Knopf.

Costanza, Robert, Bryan G. Norton, and Benjamin D. Haskell, eds. 1992. *Ecosystem Health: New Goals for Environmental Management*. Washington, D.C.: Island.

Cowell, C. Mark. 1993. "Ecological Restoration and Environmental Ethics." *Environmental Ethics* 15: 19–32.

Cronon, William. 1995. "The Trouble with Wilderness; or, Getting Back to the Wrong Nature." In *Uncommon Ground: Toward Reinventing Nature*. Ed. William Cronon. New York: Norton.

Cuomo, Chris J. 1998. *Feminism and Ecological Communities: An Ethic of Flourishing*. London: Routledge.

Denevan, William M. 1992. "The Pristine Myth: The Landscape of the Americas in 1492." *Annals of the Association of American Geographers* 8: 369–385.

Elliot, Robert. 1982. "Faking Nature." *Inquiry* 25: 81–93.

——. 1997. *Faking Nature: The Ethics of Environmental Restoration*. London: Routledge.

Evernden, Neil. 1992. *The Social Creation of Nature*. Baltimore: John Hopkins University Press.

Foote, Jeffrey P. 1973. "Wilderness––a Question of Purity." *Environmental Law* 3: 255–266.

Goodin, Robert E. 1992. *Green Political Thought*. Cambridge: Polity.

Grumbine, R. Edward. 1992. *Ghost Bears: Exploring the Biodiversity Crisis*. Washington, D.C.: Island.

——. 1994. "What Is Ecosystem Management?" *Conservation Biology* 8: 27–38.

Hargrove, Eugene C. 1989. *Foundations of Environmental Ethics*. Englewood Cliffs, N.J.: Prentice Hall.

Harper, John L. 1987. "The Heuristic Value of Ecological Restoration." In *Restoration Ecology: A Synthetic Approach to Ecological Research*. Ed. William R. Jordan III, Michael E. Gilpin, and John D. Aber. Cambridge: Cambridge University Press.

Hettinger, Ned, and Bill Throop. 1999. "Refocusing Ecocentrism: De-emphasizing Stability and Defending Wildness." *Environmental Ethics* 21: 3–21.

Holloway, Marguerite. 2000. "Nurturing Nature." In *Environmental Restoration: Ethics, Theory, and Practice*. Ed. William Throop. Amherst, N.Y.: Humanity.

Jackson, Laura L., Nikita Lopoukhine, and Deborah Hillyard. 1995. "Ecological Restoration: A Definition and Comments." *Restoration Ecology* 3: 71–75.

Jordan, William R. III. 1994. "'Sunflower Forest': Ecological Restoration as the Basis for a New Environmental Paradigm." In *Beyond Preservation: Restoring and Inventing Landscapes*. Ed. A. Dwight Baldwin Jr., Judith De Luce, and Carl Pletsch. Minneapolis: University of Minnesota Press.

Kane, G. Stanley. 1994. "Restoration or Preservation? Reflections on a Clash of Environmental Philosophies. In *Beyond Preservation: Restoring and Inventing Landscapes*. Ed. A. Dwight Baldwin Jr., Judith De Luce, and Carl Pletsch. Minneapolis: University of Minnesota Press. Kant, Immanuel. (1785) 1964. *Groundwork of the Metaphysic of Morals*. Trans. H. J. Paton. New York: Harper and Row.

Katz, Eric. 1992. "The Big Lie: The Human Restoration of Nature." *Research in Philosophy and Technology* 12: 231–241.

——. 1997. *Nature as Subject: Human Obligation and Natural Community*. Lanham, Md.: Rowman and Littlefield.

Kramer, J. Steven, and Mark Woods. 2003. "Naturalness, Wildness, Freedom: Value in a More-Than-Human World." Unpublished manuscript.

Lee, Keekok. 1994. "Awe and Humility: Intrinsic Value in Nature. Beyond an Earthbound Environmental Ethics." In *Philosophy and the Natural Environment*. Ed. Robin Attfield and Andrew Belsey. Cambridge: Cambridge University Press.

Leopold, Aldo. 1966. *A Sand County Almanac: With Essays on Conservation From Round River*. Oxford: Oxford University Press.

Lo, Yeuk-Sze. 1999. "Natural and Artifactual: Restored Nature as Subject." *Environmental Ethics* 21: 247–266.

McKibben, Bill. 1989. *The End of Nature*. New York: Random House.

Moriarty, Paul Veatch, and Mark Woods. 1997. "Hunting [NOT IDENTICAL TO] Predation. *Environmental Ethics* 4: 391–404.

Nilsson, Christer, and Gunnel Grelsson. 1995. "The Fragility of Ecosystems: A Review." *Journal of Applied Ecology* 32: 677–692.

Noss, Reed F. 1995/96. "Science Grounding Strategy." *Wild Earth* 5: 17–19.

Noss, Reed F., and Allen Y. Cooperrider. 1994. *Saving Nature's Legacy: Protecting and Restoring Biodiversity*. Washington, D.C.: Island.

Oelschlaeger, Max. 1991. *The Idea of Wilderness: From Prehistory to the Age of Ecology*. New Haven: Yale University Press.

——. 1999. "Comment: On the Conflation of Humans and Nature." *Environmental Ethics* 21: 223–224.

Pickett, S. T. A., and P. S. White, eds. 1985. *The Ecology of Natural Disturbance and Patch Dynamics*. San Diego: Academic.

Pimm, Stuart L. 1991. *The Balance of Nature? Ecological Issues in the Conservation of Species and Communities*. Chicago: University of Chicago Press.

Rolston, Holmes III. 1994. *Conserving Natural Value*. New York: Columbia University Press.

Shrader-Frechette, K. S., and E. D. McCoy. 1993. *Method in Ecology: Strategies for Conservation*. Cambridge: Cambridge University Press.

Soper, Kate. 1995. *What is Nature? Culture, Politics and the Non-Human*. Oxford: Blackwell.

Sylvan, Richard. 1994. "Mucking with Nature." *Against the Main Stream: Critical Environmental Essays*. Discussion Papers in Environmental Philosophy, no. 20. Australian National University, Canberra.

Taylor, Paul W. 1986. *Respect for Nature: A Theory of Environmental Ethics*. Princeton: Princeton University Press.

Throop, William. 1997. "The Rationale for Environmental Restoration." *The Ecological Community: Environmental Challenges for Philosophy, Politics, and Morality.* Ed Roger S. Gottlieb. New York: Routledge.

——, ed. 2000. *Environmental Restoration: Ethics, Theory, and Practice.* Amherst, N.Y.: Humanity.

Turner, Frederick. 1994. "The Invented Landscape." In *Beyond Preservation: Restoring and Inventing Landscapes.* Ed. A. Dwight Baldwin Jr., Judith De Luce, and Carl Pletsch. Minneapolis: University of Minnesota Press.

Turner, Jack. 1996. *The Abstract Wild.* Tucson: University of Arizona Press.

Walker, Brian. 1995. "Conserving Biological Diversity through Ecosystem Resilience." *Conservation Biology* 9: 747–752.

Westra, Laura. 1994. *An Environmental Proposal for Ethics: The Principle of Integrity.* Lanham, Md.: Rowman and Littlefield.

Wilson, Edward O. 1992. *The Diversity of Life.* Cambridge: Harvard University Press, Belknap.

Woods, Mark. 2001. "Wilderness." In *A Companion to Environmental Philosophy.* Ed. Dale Jamieson. Oxford: Basil Blackwell.

CONCLUSION

Autonomy, Restoration, and the Law of Nature

WILLIAM R. JORDAN III

Yet Nature is made better by no mean
But Nature makes that mean; so o'er that art,
Which you say adds to Nature, is an art
That Nature makes.

—WILLIAM SHAKESPEARE, *THE WINTER'S TALE*

I DIDN'T know quite what to make of the notion of the "autonomy of nature" when Thom Heyd invited me to contribute to this volume. I thought at first that it meant something like "independent" or "self-sufficient," ideas I have never found either appealing or coherent when applied to nature or to specific ecosystems. But then I read the book and realized this is not the case. "Autonomous" actually means something much more interesting than that, and I now see that it offers an interesting and useful way of articulating the perennial question of how to construe or construct or respond to the relationship between our own species and the rest of nature.

To begin with, the word—that "auto" right up front—asserts the selfhood of other creatures, the value that entails, and the respect due it. At the same time, unpacked from the other end, the word also raises questions about the nature of the self, pointing toward the ambiguous nature of the self and its relationships with others, providing a useful context within which to explore the problems and dilemmas posed by these relationships.

What it does *not* provide is a solution to these problems in the form of the ideal of a null relationship that tries to solve the problems—ecological, political, psychological, and spiritual—of our relationships with other subjects simply by ignoring or denying them. The word "autonomy" is useful here not because it defines an ideal but precisely because it doesn't. What it points toward, rather, is the irreducible tension that exists not just between humans and the rest of nature but between any self and any other self: the philosopher's enigma of the one and the many that, at the psychological level, we *feel* as the distance between souls. This is the insight Thoreau achieved on Mount Ktaadin, when, far from the friendly, suburban precincts of Walden Pond and appalled by the wildness he found in the forests of Maine, he perceived the otherness of nature, writing later in *The Maine Woods*: "Talk of mysteries! Think of our life in nature,—daily to be shown matter, to come in contact with it,—rocks, trees, wind on our cheeks! the *solid* earth! The *actual* world! *the common sense! Contact! Contact! Who* are we? *where* are we?"[1]

The problem, as Thoreau notes, is at bottom not really a problem at all. It is a mystery. And so the importance of ambiguity—the gateway, it may be, to mystery. And so also the difficulty—impossibility, actually—of dealing with relationship in terms of the abstractions of philosophy. These may be clear and free of ambiguity—at least relatively so—but that is exactly why they cannot ultimately come to terms with the mystery of relationship. Indeed, the sharp-edged abstractions of "nature" and "culture" leave us nowhere to go with the lives we actually live, spirit encountering matter, struggling to justify the act of ecological restoration, for example, or to clear a space for a moral agriculture from a perspective that takes the categories too seriously and provides no access to the mystery that lies *between* them.

It is there, not in the abstractions of philosophy or even the categories of thought, that we have to explore in order to deal productively with the mystery of relationship, not in the abstractions of philosophy, as theologian Catherine Pickstock has argued, but "beyond writing," in concrete experience organized by art and religion.[2]

As it happens, the word "autonomy" does accommodate this larger meaning. Even as defined in the dictionary, as Thomas Heyd notes, "autonomy" does not mean radical independence, a life that is somehow on its own, abstracted from and free of the exigencies of ecology and evolution. It means, rather, possessing a painfully ambiguous *measure* of self-rule. Thus *The American Heritage Dictionary* defines "autonomous" as "not controlled by others or by outside forces; independent . . . self-directed." Not *controlled*—of course. But certainly influenced. "Independent," certainly. But not, we must insist, in an absolute but only in a relative or *ecological* sense. This is supported by the examples provided: a nation, a corporation, a person can all have autonomy, but none of them can be independent of its surroundings. They are, in fact, in a fundamental way, actually defined by their dependencies and limitations. Ecology and evolution, of course, add their authority to this idea, taking it as axiomatic that any organism, any species, or any ecosystem is not only subject to the ineluctable influence of everything around it but is in the most fundamental sense the product, outcome, or expression of those influences at both the evolutionary or genetic and the ecological levels. This, we may note, is ultimately what it means to be a subject.

Backing up this take on the idea of autonomy is the etymology of the word, from the Greek "*autos*," or self, and "*nomos*," that is, rule or law. The idea here is that a person or thing is autonomous to the extent that it has and follows its own law. And it was realizing this that brought me around to the idea that "autonomous" might indeed be a good term to characterize a respectful relationship between ourselves and the rest of nature or among selves generally, not because it is about independence but precisely because it is, in the last analysis, about interdependence. The "autonomous" plant or animal is no more self-governing than, say, an automobile is self-mobile; both are so in striking ways but only up to a point, and in both cases such independence as there is rides on top of a whole network of dependencies.

What I mean by this is simple. At least so far as biological subjects are concerned, the "law" of their being, the genes and memes that make them what they are and give them the prerogatives and the limited degree of self-direction or freedom they enjoy, is not some kind of existential directive, downloaded from an ideal universe of Platonic forms. Neither does it lie in abstractions such as the wild or wilderness, though these have served environmental thinkers as a refuge from the ambiguity of order and hierarchy—of *nomos*—for more than a century. It is on the contrary a detailed body of legislation inscribed and worked out through the process of natural (and, no less, artificial) selection through intensive, mutually creative, ultimately life-and-death interactions with other subjects. In other words, if a plant or an animal—a starfish, say, or a walnut tree—has a certain essence, or *tele* or *nomos*, that has moral value and demands our respect, that essence is not simply about freedom. It is, rather, the "wisdom" accumulated by a species or an individual as a result of interaction with—and dependence on—other subjects, including nonliving elements such as water and sunshine. This subject, the individual or species we value, is not itself a discrete, independent entity but rather the result of countless interactions and dependencies at the ecological level transcribed into genes and at the psychological level recorded in memory and the subconscious. In other words, its "law," indeed its "nature," the basis for its autonomy, is grounded in a web of dependencies reaching back in evolutionary time that defines its freedom in concrete terms and governs its relations with other subjects in ecological time and space.

This is the point Michael Pollan dramatizes in his book *The Botany of Desire*,[3] in which he plays with the notion that domesticated species such as the potato or the tulip, far from being exploited by humans, have in fact only been exceptionally successful in putting humans to work propagating their genes.

This raises troubling chicken/egg, tail/dog questions, which may make us uneasy—exactly who is in charge here?—but which just as clearly have important implications for our relationship with other species and the rest of nature.

In fact, this question of dependence and autonomy brings us to the core of the mystery of our relationship not only with nature as other but with all others, including other people. This is, after all, not just about ourselves and the rest of nature. It is the core or nexus of an existential predicament. It is arguably what much of human anxiety is about, and arguably the primary purpose of art and religion is to articulate this anxiety and to invent ways of dealing with it in a productive way. Consider, for example, critic Frederick Garber's discussion of Thoreau's use of the figure of the creation of a clearing in the woods as a metaphor for human relations with nature, referring to what he calls "a sophisticated counterpoint of autonomy and participation" in Thoreau's reflections on the relationship between Native Americans and the forest, which he admired as nobly distanced. "It surely follows," Garber writes

> that white men will have difficulties because of their drive to cultivate; even garden plots are a violation of the integrity of man's natural partner. And there Thoreau ran into some prime difficulties. He was committed to the position that consciousness is assertive and redemptive, that is, necessarily aggressive because it is the business of mind to transform the world into the self. . . . Thoreau never could play down the importance of the activities of consciousness or his joy in indulging them. Indeed, his life was spent in refining their possibilities in a relationship between nature and the mind. In that case, how could he permit nature to be itself and yet approach it with a consciousness whose very principles see the world as something that, however delightful, has to be used by man. How would it be possible to make the mind respectful, participatory and redemptive all at once? . . . What he gathered of the Indian and what he knew of himself posed an immense challenge to the adequacy of his imagination.[4]

Generalizing this point, critic Northrup Frye argues that it is the vocation of humans to transform matter into mind and that whenever "he is doing genuine work, that is, not making war or feeding a parasitic class, he is making a human artifact out of nature."[5]

As Thoreau realized, such work is problematic. But it is not problematic because humans are in any fundamental sense outside of nature. It is problematic precisely because human beings are natural, so the difficulty here arises from our realization—Thoreau's awed response on his experience on Mount Ktaadin—that nature is familiar with alienation, is filled with subjects in ambiguous relationship with each other, is in fact a fountain or gusher of creative alienation.

Transcending the Categories

Reality, we must keep in mind, transcends the categories. Yet we live with our categories and in fact can't live without them. And what struck me in reading the preceding chapters is the hold, a century and a half after Darwin and more than a century since the emergence of ecology, the categories of "nature" and "culture," indeed the old nature/culture dualism, still has on our thinking about nature and our place in it. Both ecology and evolutionary theory offer powerful ways of thinking about nature that, in their own ways, insist on a fundamental continuity between our own species and the rest of nature. Yet in a conversation about the autonomy of nature, we find it easy to forget that the idea of autonomy itself points beyond independence to the dependencies that underlie it.

Some argue outright from the premise that nature is in essence the nonhuman, whatever is—or remains—untouched by human influence. Clearly there is value in this idea of nature, and just as clearly it is a value that is compromised—in principle irretrievably—by the human touch, however well intentioned. Yet equally clearly it is a value that, in the last analysis, nature itself denies us, at least at the level of the ecosystem and the planet, where, so far as we can tell, everything—including us—is inextricably "hitched," as John Muir said, to everything else.

At the same time, there is an important sense in which nature *is* other than culture, if not in an ultimate, ontological sense, then at least in the same, important, sense that it is something—or someone—*else*, just as dogs and cats and chickens, and for that matter other people, are someone else to *me*, and, of course, a muskrat could say the same thing. This is the basis for what Simon Hailwood has called the "otherness" idea of nature: an idea, or construction of nature, that he sees not as a basis for a dualistic philosophy of nature and culture but as the "other" that is metaphysically and psychologically prerequisite to relationship.[6]

So, the other and otherness certainly have their place in this discussion. Surely, nature-as-the-other is one of the many values we find in nature. Yet, just as surely, as Keekok Lee makes clear in her essay, "nature" has many meanings, and it is a mistake to suppose that any one of them is somehow the essential idea, the foundation on which we must build our philosophy of nature. Yet the idea persists. And it is on this ground that critics such as Eric Katz and Robert Elliott have argued against conservation practices such as ecological restoration or, having to admit in the end that they are necessary for the well-being, and even the survival, of many (and, I insist, ultimately all) of the classic landscapes that embody the nature—nature as other or given—that they seek to preserve, they allow them, but only after having denied them any validity other than that of the most onerous, uninspiring, and unrewarding of obligations, like paying off old gambling debts.

Having denied the legitimacy of this work, those taking this position not only abandon actual ecosystems to decline into an ecologically impoverished "naturalness," sacrificing them to a single, limited idea of nature. They would also deprive us of any of the pleasure or delight we might otherwise have found in the work of agriculture, of gardening or restoration, leaving us with nothing to do about nature *as* other than to establish, or declare, a kind of hermetic relationship with it, unlike anything that, so far as we know, actually exists anywhere in nature.

Obviously, this is an idea that stands little chance of widespread acceptance or application. It is important, however, because it reflects and even gives a certain authority to preservationist tendencies that, when played out in the landscape, can actually have serious consequences. These typically include the extirpation and even the extinction of species and the deterioration of actual ecosystems that regularly results from deliberate neglect in the service of a merely fragmentary idea of nature. This is the idea of nature behind Bill McKibben's widely discussed book *The End of Nature*. And it still makes itself felt in debates about strategies for the conservation of natural areas such as Chicago's Forest Preserves, the ponderosa pine forests of the southeastern United States, or the management of species such as deer in the Northeast and Midwest, often with tragic results for the ecosystems involved, as the growing frequency and ferocity of forest fires out west in recent years dramatically attest. Just as on the prairies, these fires dramatize an essential point: nature forbids disengagement and punishes the illusion of disengagement by the wholesale destruction of innocent victims. Considering all this—ecological facts that demand the attention of the moral philosopher—it should be clear that those who argue against practices such as restoration—that is, the management needed to compensate for novel or outside influences on ecosystems—bear a heavy burden. Specifically, they will have to explain how the value they seek to preserve, represented by a fragmentary, nonecological, and antievolutionary idea of nature, justifies the damage this idea causes when applied to actual ecosystems. Strikingly, none of the preceding essays addresses this issue in a systematic way.

Even those who reject this hermetic idea as the essential or defining idea of nature often find it difficult to argue forcefully against this dispiriting way of thinking. This is true even of restorationists, who often assert that the aim of restoration is the creation of a self-supporting ecosystem, despite the fact that the very notion of a self-supporting ecosystem is an oxymoron. In this volume, Ned Hettinger rightly reminds us why, citing the ecological axiom that *all* species influence their environment. Yet in clearing a space for agriculture, Hettinger is diffident, suggesting that, considering the intensity and pervasiveness of human influence on the planet today, a kind of environmental apartheid is nevertheless called for with respect to natural—as opposed to agricultural, or working—landscapes. If by this he means a commitment to minimizing human influence on certain areas, recognizing that complete isolation is impossible, that ecological influence is as pervasive and ineluctable as that of gravity and must in the last analysis always be taken into account, this is a sensible idea. But if the idea is that places really can be isolated in an ecologically effective way, that is an entirely different matter.

In a similar vein, Bill Throop and Beth Vickers offer a defense of agriculture, but only in the most conservative terms. They allow for a moral agriculture, but only if it is modeled on traditional forms of agriculture that seem, at least from the perspective offered in our own day by practices such as factory farming and modern techniques for genetic manipulation of species, ecologically rather benign. This, of course, is Wendell Berry country.[7] And though it is far from clear that it offers a socially, politically, or economically plausible scenario for the future of agriculture, if widely implemented it might actually solve certain environmental problems. Its weakness, however, is that it fails to come to terms with the creative, and necessarily manipulative, aspect of agriculture and, by extension, of technology generally. Agriculture is, not just incidentally but at its core, the manipulation of other species in a radical way, assuming control not only over their bodies and their lives but even their genes. Thus even the most benign forms of agriculture, though they may be ecologically less intrusive and perhaps more sustainable than some other forms, nevertheless represent premeditated, methodical, ongoing violence against other species and ecosystems. Even if you solve the ecological problem of exploitation beyond the carrying capacity of the system, this more fundamental, psychological problem of manipulation remains, so that a philosophy of relationship with nature has either to find some way of coming to terms with this violence or repudiate agriculture or, short of that, confine agriculture to certain, "working" landscapes, keeping other, "natural" landscapes wild, free, "autonomous"—and ecologically doomed.

A Tale of Two Prairies

Autonomy, as I pointed out earlier, is a useful idea here not because it solves the problem of our place in the world but because, fully understood, it *defines* the problem: what *is* the autonomy of the other creature, species, or landscape? What is the "law" that defines that particular other and makes it who it is, not in large, abstract, and categorical terms but exactly and specifically, in ecological detail? And how can we best respect the "law" or nature that defines the particular terms of that autonomy, keeping in mind that it is itself the outcome and reflection of a history of ecological and evolutionary interaction and dependence?

As Lee and others point out in the earlier chapters in this book, the challenge we face is not simply to respond to one piece of a deeply ambiguous idea but to find a basis for relationship with other creatures that somehow accommodates the mutually contradictory facts of our experience that are reflected in the many meanings of "nature." It is true that this places us in an uncomfortable position. But that, I will argue, is precisely its greatest value. We *belong* in an uncomfortable position, not because we are outside nature or exceptional to it but because nature itself is uncomfortable—that is what creation is all about—and because we are so deeply natural.

I see three problems with the idea of nature, disconnected from its contradictory opposite, that is articulated by Katz and supported in various ways by Dean Bavington and John Sandlos in their essays. The first is that, as I noted earlier, it is antievolutionary. It is antievolutionary because it overlooks the idea of "natures" as both constructed and expressed through interaction and dependency, an idea that has been foundational for biology since Darwin. And it is antievolutionary—in fact is specifically creationist—because it locates the value of a thing in its origin rather than in its participation in an ongoing creation. Besides this, it is antiecological, since it simply ignores the unavoidable influence humans (like any species) have on ecosystems. And it is, perhaps most obviously, antihuman because it has nothing to say about the existential crisis our species experiences in its necessarily violent participation in creation.

Since I have already touched on the point about evolution, let's take up these last two points one at a time, beginning with the question of ecology and ecological relationships. The interconnectedness of things is axiomatic in ecology, as in evolutionary theory, and environmentalists often appeal to it when making the case for human dependence on nonhuman nature. They generally have less to say about nature's dependence on us. But, of course, it works both ways. It is true that we depend on nature, but it is just as true that nature—at least in the concrete form of actual ecosystems—reflects the influence of humans just as it reflects the influence of other species and so in a crucial sense depends on them. Though it is certainly true that the interaction varies widely in both nature and intensity and that the coefficient of interaction is sometimes very small, it is, axiomatically, never zero. There is, neither in principle nor in fact, any way to isolate an ecosystem from everything else around it, no fence so high, no pickle jar or thermos bottle so exclusive that it can protect any ecosystem anywhere from the influence of humans or any other species inhabiting this planet. Pretending otherwise, indulging in the fantasy of preservation as a means rather than as the goal it properly is, is simply to abandon the historic ecosystem to drift in the variable wind of influences arbitrarily defined as coming from outside the system.

To make this point, I often offer the example of two ecological communities in the collection of restored communities maintained by the University of Wisconsin Arboretum in Madison, where I worked for twenty-four years. This is what I call my tale of two prairies. There are at the Arboretum two prairies. One occupies a site on the slope and crown of a hill where the soil is thin, and the several generations of farmers who worked the land there never broke ground. Leaving it alone, they in effect preserved it. But, of course, in the absence of fire, kept away by the breaking of land all around it, this site rapidly grew up into a grove of trees. Today, it is an oak woods—long ago named Noe Woods—dominated by the 150-year-old oaks that date back to the last prairie fire here. What was a tallgrass prairie is now a forest, and virtually all the prairie species are gone.

The story is quite different on the level ground that stretches for a half mile or so to the east. Here, settlers broke the prairie and cultivated the land for the half century or so straddling the turn of the twentieth century. When the university acquired that parcel of land in 1932, most of it was pasture and old field, and here, as in Noe Woods, few of the native prairie species remain. On this site, however, in 1936 Civilian Conservation Corps crews working under the supervision of Theodore Sperry began one of the earliest attempts at large-scale prairie restoration. These proved reasonably successful, and as a result this piece of ground, amounting to about sixty acres, is now actually a prairie. It is true that it is an artifact and that it is ecologically and historically discontinuous from the old, or "original," prairie. It harbors more than three hundred species of prairie grasses and forbs, however, as well as a considerable number of prairie birds and mammals, providing a regional hot spot of native biodiversity *and* a reasonably accurate swatch—facsimile, copy, fake, whatever you choose to call it—of one of the richest, and now one of the rarest and most diminished, of all the classic ecological communities of North America.

Noe Woods, of course, is the preserved prairie, the beneficiary—or victim—of neglect and of treatment that at least roughly approximated the hands-off management Katz prescribes. Never subject to the ineptitude and disgrace of human administration or the danger of iatrogenic damage decried by Dean Bavington, it is, we may say, free, "autonomous," and wild. It is also no longer a prairie. It is in fact the stretch of land to the east, where the old, "natural" prairie was destroyed and then brought back by a painstaking, deliberate, groping, purposeful effort, that is actually a prairie—named John

T. Curtis Prairie, after the ecologist who participated in its creation. That effort, now extending over almost three-quarters of a century, certainly entailed many mistakes. Some of these were a result of insufficient knowledge, others of administrative glitches, limited budgets, or even mistakes, such as mistaking the identify of a plant. There is always a risk of iatrogenic harm, of course, but that is a risk that in the extremity of life—of a person or an ecosystem—we may be morally obliged to take. The classic injunction to the physician to "first, do no harm" always has to be accompanied by a willingness to risk harm, nevertheless. And mistakes do happen. But fortunately physicians—and restorationists—learn from their mistakes.[8]

Some, I realize, will object to my tale of two prairies. They will point out that this is only one ecosystem, that prairies are not only fire-dependent but respond to changes in fire frequency much more rapidly than some other fire-dependent communities such as the slash pine forests of the southeastern United States or the ponderosa pine forests of the Southwest. This, however, is missing the point. Certainly, it is true that the prairies are unusually responsive in this way; they are themselves, we now realize, partly the creation of pre-Columbian peoples, and it happens that, for purely ecological reasons, they respond to changes in fire frequency relatively quickly and dramatically. But this does not make them exceptional in any fundamental way. It simply makes them a kind of indicator ecosystem, excellent occasions for learning a lesson that in the last analysis applies to every square foot of the planet: *there is simply no way to disengage*, no way to have no influence on the plants and animals on the other side of the fence. The question this raises is, how much of the natural landscape are we willing to sacrifice for the sake of a single part of the rich and complex idea of "nature"? What is at stake, after all, is not trivial. It is nothing less than the survival of an entire ecosystem—not just Curtis Prairie but *all* the prairie, and not just prairie but ultimately every ecosystem on the planet. Unless we reject outright the axiom of ecological connectedness, the conclusion is unavoidable: Curtis Prairie is not exceptional. It is paradigmatic. What has happened there defines, like it or not, the terms on which the "natural" landscapes and ecosystems of the future will exist. And this is the key not only to their survival but to their autonomy, as I shall argue below.

From "Autonomy" to Specific Autonomies

How, then, can we respect the autonomy—the law or nature—of the other without violating that other law of nature—the law that forbids disengagement?

There are two questions here. The first is how can we come to know the nature of the other subject, so that we can respect it. As I argued above, it turns out that restoration is one of the most powerful ways we have of doing this, since it provides a way of testing our ideas against the reality of actual ecosystems. Thus it turns out that the best way to respect the autonomy of a tallgrass prairie is to burn it at irregular intervals, and this is a lesson we owe in large part to the work of restorationists.

The second question is, what can we do about the crisis of conscience and of sensibility that we experience when we confront the limits of autonomy, in the act of killing, for example, or manipulation of other species? "See, at some point we have to connect with the rest of nature," literary critic Frederick Turner has written, "and it always involves death."[9]

Hunters know this, encountering the travail of creation in the act of killing in order to sustain life. But farmers experience the at-oddsness of creation even more intensively because they participate more intimately in the process, officiating, so to speak, not only at the death but also at the birth of the prey plant or animal,[10] replacing the hot-blooded killing of the hunt with the cold-blooded betrayal and premeditated murder of the barnyard. (It is, of course, no help here to identify the other species as "subjects." That is not the solution to the problem but in fact only intensifies it. In fact, it *is* the problem, since now what you are killing is your brother or your sister or an ancestor, a subject, or even a person, with prerogatives and moral value.)

In either case, something must be done in recompense for this crime: the crime of creation and the radical destruction it entails. And in either case it is clear that literal recompense is impossible: death is irreversible, a dramatic encounter with the tragic one-wayness of time. Life must be taken to feed life and, once taken, cannot be given back. The problem, in fact, transcends moral cost-accounting. It lies beyond, or behind or beneath, the calculations of ethics, the ways we calculate what is right and wrong: this cannot be right and cannot be made right. And so, to deal with this, humans, hunters as well as pastoralists or farmers, have dealt with it by stepping out of the literal and the logical and into the dimension of the imagination, using for this the technologies of the imagination, performance in the form of story, myth, dramatic enactment and reenactment, and ritual.[11] Indeed, it is arguably the discovery that this can work, that the god, so to speak, will settle for partial payment of the debt owed, the cost incurred simply by living, that makes a psychologically and morally coherent relationship with others possible at all.

Thus hunters develop elaborate rituals characteristically entailing measures that ensure the recycling of the soul of the slain animal. Social philosopher Jonathan Z. Smith has argued that sacrifice is a kind of miniaturization of the process of domestication, invented by agricultural peoples.[12] And Timothy Ingold has developed a similar set of ideas, characterizing sacrifice as part of the process by which herding peoples justify their appropriation or taking charge of souls previously under the spiritual/political sway of animal-master spirits[13]. To this, I add the suggestion that this strategy works because it provides a way of concentrating the horror and shame that is inseparable from agriculture, making it possible to focus on it and deal with it productively.[14]

What is important here is not that rituals such as sacrifice somehow justify the taking of life. That can neither be justified nor avoided; that indeed is our existential predicament. What is important is that they reflect an awareness of the trouble that is at the heart of creation—we might as well say "nature"—and of human complicity in this trouble, that they offer a way of focusing and articulating this experience—what mythologist Joseph Campbell called "the qualm before the deed of life"[15]—and then dealing with it in a psychologically productive way, and that the result of all this is not denial or dismissal or detachment but deeper awareness and values such as community and beauty. (Thus, in Christian tradition, to take a convenient example, it is the ritual murder of God in the Eucharist that is the paradigmatic act of communion with the divine.)

From this perspective, certain things fall into place, helping us to make useful distinctions regarding the question of autonomy. John Sandlos, for example, argues that the restorationist's attempt to remove exotic species from an ecosystem reflects an unhealthy puritanism, a desire to cleanse the ecosystem of elements that have been, perhaps arbitrarily, defined as extraneous, as a kind of metaphysical "dirt" or weed. Sandlos cites anthropologist Mary Douglas's discussion, in her classic *Purity and Danger*, of this kind of assertion of categories as "an instrument of control" over others, and, of course, it can be. Yet what society can exist without a shared set of categories with which to make sense of the world, dangerous as that may be? Sandlos sees restoration, with its preoccupation with the exclusion, elimination, or control of exotic species, as a kind of puritanism, and my sense is that that may be true at times, as, for example, when attempts to remove an exotic such as buckthorn from oak forests and savannas in the Midwest or tamarisk from arid areas in the West take on the spirit of a crusade against the unwanted other. On the other hand, at a more fundamental level, restoration is itself a confrontation with the other and, more than that, a violation of categories, specifically those of "nature" and "culture." That is, no doubt, why environmentalists ignored and resisted it for most of the past century. And this is a matter of considerable moral consequence in light of Douglas's argument that it is not by discarding categories but by *retaining* them and then ritually violating them that people achieve transcendence and access to higher values such as community and the sacred.[16]

Similarly, Throop and Vickers object to the spectacle of a dancing bear, seeing this as a violation of the bear's autonomy. I am inclined to share this view, though I have to admit that my notion of bear autonomy reflects only a negligible, second-hand knowledge of bear nature, and I would quickly defer here to the judgment of anyone who has cultivated a serious relationship with bears. On the other hand, I would suggest that other kinds of interaction with other animals, including rough ones, as in rodeo, for example, or even violent ones, as in bullfighting, may be entirely in keeping with the autonomy of the other animal, since they act out relationships that are, we have good reason to suppose, part and parcel of what it means to be a steer or a bull.[17]

If it is possible to spin value out of the natural crime of agriculture in this way—and human cultures arguably rest at the most fundamental level on precisely these acts of psychological and spiritual jujitsu—then it stands to reason that it might be possible to do something of the kind with respect to an activity such as ecological restoration. Restoration, after all, is, in a purely technical sense, a form of agriculture and so is, as Professor Katz notes, part of the human project. But it is also, so to speak, agriculture in reverse, an attempt not to take and transform but to give back and reconstitute. While there is no denying that it is intentional and deliberate and reflects human perspectives and even desires, it is important to understand that in its purest, or most ambitious form it represents an exercise in the *giving up of deliberation* and even of desire in the attempt merely and blankly to copy—not creatively to imitate—but to *copy* "nature" in its aspect as the given or merely encountered.

It is an exercise in autonomy that leads beyond the idea of autonomy as wholly and simply free to a richer idea of autonomy as the product of engagement, dependency, and the exchange of goods and services. It is also an exercise that allows us to explore in a very discerning way what we might call the limits or definition of the autonomy of the subjects we are working with, whether we see this as an ecosystem such as a prairie or wetland or the various plants and animals that make it up.

We can say this because the core idea of restoration is to *let alone*, to free the subject from our influence, to behave in such a way that we become ecologically invisible to that subject, so that it can continue to behave or resume behaving as if, in fact, we were absent or were not influencing the subject, while at the same time it acknowledges—and indeed dramatizes in the language of nature, the language of action and the exchange of goods and services—the irreducible fact that we *are* there and *are* influencing the subject.

These are the questions the restorationist is always asking: How little can I get away with? When can I stop? When will the ecosystem take over for itself, and in just what ways? And it is just this that makes restoration an exploration of autonomy that leads beyond the abstract idea of autonomy as merely wild or free to the richer, more particular, more grounded idea of the autonomy of a particular subject, itself the outcome and expression of a long evolutionary and ecological history of interactions and dependencies.

This exercise, the groping toward an ideal, *absolute* autonomy for the ecosystem that lies behind the work of restoration, is, I argue, the best possible thing we can do to respect the actual, *relative*, ecological and evolutionary autonomy of other organisms at the community and ecosystem levels. It is, at the ecosystem level, the same thing we aspire to do in other areas, such as medicine, for example, or education, where the aim is always to interact with the subject in such a way as to enhance its selfhood: to heal as well as cure, to educate as well as inform, instruct, and inculcate.

And it is, of course, an *exercise*, in the sense of an exercise in music or a spiritual exercise. It is a trial run and an experiment, and it is never fully successful. But this brings us to Andrew Light's provocative suggestion that, though restoration in the fullest and strictest sense may indeed be impossible (as it is impossible for the hunter to restore the life of his prey or the farmer to reconstitute the purity and givenness of the plant or animal she has domesticated), we may be called on to make the attempt, and that may be the proper moral response to a creation in which we are, like it or not, complicit and that proceeds by making ineradicable, morally unjustifiable changes.

This works, however, only if we are careful about what we mean by "restoration." That is, it works only for restoration defined in a particular, very narrow way. For many years, I have insisted on a conception of restoration that many of my colleagues in the restoration community regard as extreme, unnecessarily narrow, and even culturally exclusionary. This is the idea of restoration defined as what Marc Hall and I recently termed "holistic restoration": the attempt to return an ecosystem or landscape to some previous condition, including all its features, and *with a studied indifference to human interests.*[18]

I am aware that, while we may try to be indifferent to our own interests in deference to a particular landscape or to the idea of nature as given, we never fully succeed at that. I am also aware that there are many sorts of restoration, depending on how you define the thing you set out to restore. We can, for example, broaden the target by setting out to restore a single species, as wolves have recently been reintroduced in Yellowstone Park or wild turkeys have been reintroduced in some areas in the Midwest in recent decades. Or we can set out to restore a single process, such as nitrogen cycling or fire or intermittent flooding. Or we can even aim for an abstract feature such as freedom or wildness, as Mark Woods suggests in his discussion of the restoration of the Everglades.

Defining goals in this rather tolerant way makes restoration easier. But it also deprives it of the value it might otherwise have of *being* impossible and therefore providing an occasion for encountering the irreversibility of time and of change, including the consequences of one creature imposing on, exploiting, or even controlling another. That, I have insisted, comes through most clearly when we set out to do the impossible: to reverse time and to restore to other species or to whole ecosystems their "original" autonomy.

This is not, I should stress, a notion of restoration that is widely shared, even among restorationists. I have insisted that it is important, however, not as *the* definition of restoration but as a form of restorative land management that generates its own distinctive kind of value, precisely because I see restoration in this sense as the best possible response to the question of what to do about nature in its aspect as given or other. It is—and this point is critical—a very different thing indeed from the managerial, regulation-driven, manipulative activity decried by Eric Katz and Dean Bavington, with respect to the morally crucial matter of intent. In fact, it is, psychologically speaking, its polar opposite. And I do not think this idea of restoration, this sense of what it is about, this commitment to copy the given in the spirit of what Robert Frost called a "tribute of the current to the source,"[19] is merely a figment of my imagination. Restoration may be, technically, a form of agriculture or gardening. But restorationists practicing holistic restoration clearly experience it as very different from traditional forms of agriculture, which are essentially creative in their dealings with nature. Restoration, in the sense I am using the term, forswears creativity in the commitment, as I said earlier, merely to copy.

Copy. Copy. We hate that word, because we make a fetish of creativity. Even restorationists object to my insistence on this, and, of course, it is true that the actual work of restoration can be highly creative. Yet there is a crucial sense in which it is at bottom noncreative, an act of deference to nature, an attempt to give back in kind what we have taken from nature. This, as I have argued, has a value of its own, not offered by any of the other games we play with nature. And, in fact, restorationists themselves make it clear that the experience of restoration is very different from the experience of gardening. "I am not a gardener," Chicago-area restorationist Bob Betz will insist, standing in the prairie he has weeded and planted for two decades. "Oh, I'd never do that," another will say. "That's just gardening."[20]

The difference, of course, is in the intent, in the restorationist's intention to do what has—*has*—to be done to ensure the survival and well-being of the system, while at the same time *not* controlling it, *not* violating its autonomy, but rather turning it back into itself, into its "original" freedom and wildness. This, as Katz and others will be quick to point out, never quite works. And, in fact, the result is a different thing—even in many important ways a different kind of thing—from the original. But since when did nature, in all the cataclysmic rush of creation, promise that anything would remain the same?

What restorationists are learning is that, in fact, it is possible to have the old thing, or at any rate a pretty good copy of it, but—naturally—only on new terms. The new terms reflect new conditions, in most cases the role humans have played in the shaping of the landscape. As an attempt to reverse or compensate for changes we have brought about, restoration turns out to be a reflection on those changes, bringing them into human awareness. At the same time, as a powerful way of learning and testing ideas about the landscape and its history, restoration, properly carried out, generates knowledge about it, inscribing, as it were information from the old form, coded in genes, to the new form, represented by human consciousness. This uploading results in a new kind of entity, a new autonomy, that incorporates the nonhuman into the human in a new way, ensuring, or at least making possible, the survival of the old species in the new context.

Far from depriving other species of a radical freedom they never had in the first place, this whole process enlarges their freedom by linking it with that of the freest creature we know: ourselves. As I have suggested elsewhere, it confers on these species and the ecosystems they inhabit the prospect of survival on the new terms of creation created by nature itself. It confers nothing less than a kind of relative, or ecological, immortality.[21]

And that, I suggest, is as good as it is going to get.

Of course, this is not without regret, without some version of what Mircea Eliade called nostalgia for paradise or existential dismay on encountering what Frost called "the stream of everything that runs away." But it is, as Frost suggested, precisely the regret that is sacred. Wildness in the comfortable old sense will always have its attractions. But the real wildness will always be the wildness of the really new thing. And it is not, after all, really in short supply. It is only in short supply in the places where we live. Everywhere else, at the frontiers of human contact, perhaps along the continental shelves or in the mountains, certainly from a few hundred miles up and on from there through much of the universe, it still prevails. If this is not convenient, it is worth keeping in mind that, by nature, it never was.

NOTES

1. Henry David Thoreau, "Ktaadin," in *Thoreau's Complete Works: The Maine Woods, Cape Cod and Miscellanies* (Boston: Houghton Mifflin, 1929), 79.
2. Catherine Pickstock, *Beyond Writing: The Liturgical Consummation of Philosophy* (Oxford: Blackwell, 1998).
3. Michael Pollan. *The Botany of Desire: A Plant's-Eye View of the World* (New York: Random House, 2002).
4. Frederick Garber. *Thoreau's Redemptive Imagination* (New York: New York University Press, 1977), 34–35.
5. Northrup Frye. *Creation and Recreation* (Toronto: University of Toronto Press, 1980), 21.

6. Simon Hailwood. *How to Be a Green Liberal: Nature, Value and Liberal Philosophy.* (Montreal: McGill–Queens University Press, 2004).
7. Poet/philosopher/farmer Wendell Berry has written a great deal on the ecology and psychology of agriculture over the years. See, for example, his *The Unsettling of America: Culture and Agriculture* (New York: Avon, 1977).
8. The value of restoration as a technique for basic research—a way of systematically learning from mistakes—is explored in some detail in William R. Jordan III, Michael E. Gilpin, and John A. Aber, *Restoration Ecology: A Synthetic Approach to Ecological Research* (Cambridge: Cambridge University Press, 1987).
9. Frederick Turner, "She's Come for an Abortion: What Do You Say?" *Harper's Magazine* (November 1992): 53–54.
10. Jean Pierre Vernant, *Myth and Thought Among the Greeks* (London: Routledge and Kegan Paul, 1983). Professor Hunnicutt teaches in the Department of History at the University of Iowa.
11. I am making use here of ideas developed by Frederick Turner, summarized in his *The Culture of Hope: A New Birth of the Classical Spirit* (New York: Free, 1995). What is involved here is what theologian Catherine Pickstock has called a move beyond philosophy into the realm of ritual. For her discussion of this in the context of an exhaustive discussion of the Mass as ritual, see her *Beyond Writing.*
12. Jonathan Z. Smith, "The Domestication of Sacrifice," in Robert G. Hamerton-Kelly, ed., *Violent Origins: Ritual Killing and Cultural Formation* (Stanford, Calif.: Stanford University Press, 1987), 191–238.
13. Tim Ingold, *The Appropriation of Nature: Essays on Human Ecology and Social Relations* (Manchester: Manchester University Press, 1986). Yi-Fu Tuan offers a brief overview of related ideas in his *Morality and Imagination: Paradoxes of Progress* (Madison: University of Wisconsin Press, 1989).
14. William R. Jordan III, *The Sunflower Forest: Ecological Restoration and the New Communion with Nature* (Berkeley: University of California Press, 2003), 39–40.
15. Joseph Campbell. *The Masks of God: Primitive Mythology* (New York: Penguin, 1959), 180.
16. Jordan, *The Sunflower Forest,* 126–28.
17. Though certainly no authority on rodeo, I recall in this connection attending one in Denton, Texas, with the environmental philosopher Susan Bratton some years ago and our sharing the impression that the bulls, in particular, were on to the whole affair and perhaps even relished the opportunity to, as Professor Bratton put it, stomp on a cowboy in the context of a work of performance art.
18. William R. Hordan III and Marcus Hall, "Ecological Restoration," in Shepherd Krech III, J. R. McNeill, and Carolyn Merchant, eds., *Encyclopedia of World Environmental History* (New York: Routledge, 2004), 1:371–78. See also Jordan, *The Sunflower Forest,* 21–22.
19. In his poem "West-Running Brook."
20. Jordan, *The Sunflower Forest,* 87.
21. Ibid., 133–36.

Contributors

DEAN BAVINGTON, from St. John's, Newfoundland, is a B.Sc.(Hon.), MES, and Ph.D. (abd) in geography and environmental studies at Wilfrid Laurier University (Waterloo, Canada). His research focuses on a critical examination of managerial ecology and the history of managerial interventions in the cod fisheries of Newfoundland and Labrador. Awards include an O.A.S. Fellowship at the University of Northern Arizona and a Fulbright Scholarship at the University of California, Berkeley. Recent publications include "The Feminist Political Ecology of Fishing Down: Reflections from Newfoundland and Labrador," in Studies in Political Economy (2004); "Managerial Ecology and Its Discontents: Exploring the Complexities of Control, Careful Use and Coping in Resource and Environmental Management," in Environments (2002); and "From Jigging to Farming," in Alternatives (2001).

NED HETTINGER teaches at the College of Charleston in South Carolina (USA), where he is professor of philosophy and coordinator of the minor in environmental studies. He has published several dozen papers, including articles on intellectual property in Philosophy and Public Affairs (1989), on biotechnology in the Boston College Environmental Affairs Law Review (1995), on disequilibrium ecology and wildness value in Environmental Ethics (1999), on exotic species in Environmental Values (2001), on environmental disobedience in A Companion to Environmental Philosophy (Blackwell, 2001), and on a positive role for humans in nature in Ethics and the Environment (2002). His articles have been reprinted over a dozen times. "Justifying Intellectual Property," Philosophy and Public Affairs (1989) was reprinted in The Philosophers Annual, edited by P. Grim, G. Mar, and P. Williams (Ridgeview, 1991), as one of the ten best philosophical articles written in 1989.

THOMAS HEYD teaches in philosophy at the University of Victoria (Canada). He writes in environmental ethics, history of philosophy (seventeenth century), aesthetics of rupestrian art, and aesthetics and environment. His publications include Environmental Consciousness and Action (coauthor with Wolf Schluchter, German Environment Ministry, 1996), Aesthetics and Rock Art (coeditor with John Clegg, Ashgate, 2005), chapters in books, and articles in journals such as Environmental Ethics, Environmental Values, Journal of Human Ecology, British Journal of

Aesthetics, Journal of Aesthetics and Art Criticism, and Philosophy East and West. Presently he is preparing a collection of essays in environmental philosophy titled After Nature: Encountering Nature in Hybrid Spaces for Ashgate Publishing.

DR. WILLIAM R. JORDAN III is director of the New Academy for Nature and Culture (Woodstock, Illinois, USA), and codirector of the Institute for Nature and Culture at DePaul University. Widely recognized for his contributions to thinking about ecological restoration, he was the founding editor of Ecological Restoration, the first journal in this field, and a founding member of the Society for Ecological Restoration International. His book The Sunflower Forest: Ecological Restoration and the New Communion with Nature was published by the University of California Press in 2003.

ERIC KATZ (B.A. Philosophy, Yale; Ph.D., Boston University) is professor of philosophy and director of the Science, Technology, and Society Program at the New Jersey Institute of Technology (Newark, New Jersey, USA). He is the author of Nature as Subject: Human Obligation and Natural Community (Rowman and Littlefield, 1997) and has coedited (with Andrew Light) the collection Environmental Pragmatism (Routledge, 1996); (with Andrew Light and David Rothenberg) the collection Beneath the Surface: Critical Essays in the Philosophy of Deep Ecology (MIT Press, 2000); and (with Andrew Light and William Thompson) the textbook Controlling Technology, second edition (Prometheus, 2003).

KEEKOK LEE is honorary chair in philosophy of the Institute for Environment, Philosophy and Public Policy, University of Lancaster, United Kingdom. Her recent current research/publication interests include environmental philosophy, philosophy of technology, and their links from the ontological standpoint. Her latest major publication is Philosophy and Revolutions in Genetics: Deep Science and Deep Technology (Palgrave/Macmillan, 2002). She has recently been appointed to the UK Agriculture and Environment Biotechnology Commission.

ANDREW LIGHT is assistant professor of environmental philosophy and director of the environmental studies program at New York University (New York, USA). He is also a research fellow at the Institute for Environment, Philosophy and Public Policy at Lancaster University (UK). Light is the author of Reel Arguments: Film, Philosophy, and Social Criticism (Westview, 2003) and has edited or coedited fourteen books on environmental ethics, philosophy of technology, and aesthetics, including Environmental Pragmatism (Routledge, 1996), Technology and the Good Life? (coedited with Eric Higgs and David Strong; University of Chicago, 2000), Moral and Political Reasoning in Environmental Practice (MIT Press, 2003), and The Aesthetics of Everyday Life (Columbia University Press, 2005).

VAL PLUMWOOD, Australian Research Council Fellow at the Australian National University (Canberra, Australia), has published four books and over one hundred papers, mostly in environmental philosophy. Plumwood brings both modern and ancient philosophy and feminist theory to her work on revising our conceptions of human identity and interspecies relations; "shaking philosophy to its foundations," said one reviewer of Val Plumwood's ecofeminist classic Feminism and the Mastery of Nature (Routledge, 1993). A critic of anthropocentrism since 1975, Val

Plumwood's newest book is Environmental Culture: The Ecological Crisis of Reason (Routledge, 2002). A recent essay on Plumwood's work can be found in 50 Key Thinkers on the Environment, ed. Joy A. Palmer (Routledge, 2001), 283–290.

JOHN SANDLOS recently completed his doctorate in the faculty of environmental studies, York University (Toronto, Canada). His dissertation examines the impact of wildlife conservation on Aboriginal hunters in the Northwest Territories (Canada). He is presently holding a postdoctoral fellowship at the University of Western Ontario (London, Canada) funded by the Social Science and Humanities Research Council of Canada. His most recent publications include "Landscaping Desire: Poetics, Politics and the Early Biological Surveys of the Canadian North," Space and Culture (2003), and "Where the Scientists Roam: Ecology, Management and Bison in Northern Canada," Journal of Canadian Studies (summer 2002).

DR. WILLIAM THROOP is provost and professor of philosophy and environmental studies at Green Mountain College (Poultney, Vermont, USA). He has published articles in environmental ethics and epistemology, and he recently edited Environmental Restoration: Ethics, Theory and Practice (Humanity Books, Prometheus Press, 2000) and coedited, with J. Arthur and A. Shapiro, Reason and Culture: An Introduction to Philosophy (Prentice-Hall, 2001). His current research projects are focused on value conflicts that arise in restoration practice and on the challenges of educating citizens for a sustainable society.

BETH VICKERS lives in Danby, Vermont (USA). She received her B.A. magna cum laude from Green Mountain College in May 2001 with a self-designed major in sustainable community development. While at Green Mountain, she was a member of Alpha Chi and the College Honors Program. In November 2002, she returned to the college as a VISTA worker responsible for developing community outreach and service-learning programs. In this capacity, she has been working actively with surrounding communities in Vermont to promote sustainable development. She contributed to the paper in this volume as a research assistant to William Throop. She is also a professional cook.

MARK WOODS is associate professor in the Department of Philosophy at the University of San Diego (USA). His area of teaching and research specialty is environmental philosophy, and he currently is completing a book on the concept of wilderness and the philosophy of wilderness preservation. He worked as an environmental ethics consultant with the Southwest Florida Water Management District, a state agency responsible for managing surface and groundwater in south central Florida, where he facilitated community meetings between water-conflicted groups of people and developed a middle school curriculum on water management and environmental ethics.

Index